THE WEST HIGHLANDS
AND THE HEBRIDES

THE WEST HIGHLANDS
and
THE HEBRIDES

A Geologist's Guide for Amateurs

BY

ALFRED HARKER

CAMBRIDGE

At the University Press

1941

CAMBRIDGE
UNIVERSITY PRESS

University Printing House, Cambridge CB2 8BS, United Kingdom

Cambridge University Press is part of the University of Cambridge.

It furthers the University's mission by disseminating knowledge in the pursuit of
education, learning and research at the highest international levels of excellence.

www.cambridge.org
Information on this title: www.cambridge.org/9781107536777

© Cambridge University Press 1941

First published 1941
First paperback edition 2015

A catalogue record for this publication is available from the British Library

ISBN 978-1-107-53677-7 Paperback

Cambridge University Press has no responsibility for the persistence or accuracy
of URLs for external or third-party internet websites referred to in this publication,
and does not guarantee that any content on such websites is, or will remain,
accurate or appropriate.

EDITOR'S PREFACE

The lure of Scotland's western seaboard has been a theme for many pens. Its long, indented coast-line, its chains of islands, its lofty peaks and serrated ridges, its bold promontories and winding fiords possess an irresistible fascination. To the geologist it is a region of surpassing interest. The results of years of detailed research within its borders have not only added enormously to our knowledge of tectonic and igneous processes but have in many ways deeply influenced the course of geological thought. A simple exposition, however, of its beauties of form in relation to the framework of rock has hitherto remained undevised.

Dr Harker's early work among the hills of Skye and the Small Isles of Inverness-shire was an inspiring contribution to the elucidation of the complex history of the Tertiary volcanic centres. He brought a new outlook to the problems of igneous geology and, both in this country and overseas, became a recognized leader in this branch of research. In his later years he returned to the Western Highlands as an interpreter of their scenery and structure to a wider public, and the present volume is the result. In it the reader will find the scenery of this region, with its well-nigh endless variety of configuration, brought before him in a series of drawings, simple in outline but effectively portraying the relationship that exists between the features of a landscape and the rocks of which it is composed. Mountains and glens, sea-lochs, headlands and islands, follow one another in procession, forming an introduction to the science of geology itself from an area where almost every type of rock structure is displayed.

The sketches and brief descriptions which accompany them will be welcomed by geologists and geographers alike. But the book will, it is certain, have a much more general appeal. It will be especially valuable to those whose journeyings take them across the narrow seas or round the islands of our western coast, and many a striking panorama will be found depicted and described in its pages. Its aim is threefold: to epitomize the geology; to interpret the scenery; and to arouse in the traveller something of the eager, inquiring spirit of the author himself.

The circumstances in which the book is published and the condition of the manuscript left by Dr Harker are stated below in a contribution by Professor J. S. Boys Smith, one of Dr Harker's executors, with whom the initiative has lain for interpreting the author's partly expressed wishes, and a biographical sketch is added by his life-long friend, Sir Albert Seward. Professor C. E. Tilley, a former pupil of Dr Harker and now head of the Department with which Dr Harker was so intimately connected, has also throughout been closely associated with the arrangement and completion of the book. At their invitation, and with their constant help, the final preparation of the manuscript and illustrations for publication has been carried out, and the knowledge gained through their personal association with Dr Harker has been freely utilized. In the completion of certain sections left unwritten by Dr Harker, assistance has been given by geologists who have a special acquaintance with the districts concerned, and for this and other help especial thanks are due, in addition to Professor C. E. Tilley, to Dr E. B. Bailey, Dr M. Macgregor and Mr D. Haldane, Geological Survey, to Dr R. M. Craig, Edinburgh University, and to Mr J. Mathieson, Royal Scottish Geographical Society. The parts which have been added to the original manuscript are enclosed within square brackets. A table of geological formations and a glossary of technical terms have also been supplied.

The geological map (I) and topographical maps (V–VIII) have been compiled and drawn in Edinburgh, and for supervising this work many thanks are accorded to my colleague, Mr G. G. Torkington. The small geological maps (II–IV) are reproduced from previous publications, and for permission to use them we are indebted to the Council of the Geological Society of Glasgow (Map IV) and to the Controller, H.M. Stationery Office (Maps II and III).

J. E. RICHEY
H.M. Geological Survey,
Edinburgh

August 1940

CONTENTS

CONTENTS

ILLUSTRATIONS

Alfred Harker in 1936, from a photograph   *facing page* xvii

FIGURES

ILLUSTRATIONS

ILLUSTRATIONS

MAPS

NOTE ON THE MANUSCRIPT LEFT BY ALFRED HARKER AND ON ITS PREPARATION FOR PUBLICATION

Alfred Harker, at his death on 28 July 1939, left the manuscript of this book uncompleted, but in an advanced state of preparation. Six weeks before he died, when his strength was already failing, he told me that he was writing a small book on the West Highlands, to be illustrated by his own sketches, but that he feared he might not live to complete it. He intended it for travellers and others interested in geology and in scenery. He also mentioned the book in general terms to Professor C. E. Tilley on more than one occasion. He spoke of it little, if at all, besides; and he left no directions with regard to the manuscript. It fell to me as an Executor, to whom he had entrusted the care of his books and papers, to collect the material on which he was working. The advanced state of the book, and his known intention to publish it if he should live to complete it, appeared to justify, and indeed to demand, the publication of what thus becomes the last of his books. The ready co-operation of Sir Albert Seward and Professor C. E. Tilley, and the help of Dr J. E. Richey of H.M. Geological Survey, have enabled the book to appear. Dr Richey, who contributes the Preface, very generously undertook to prepare the manuscript for publication and to provide such additions as were needed to complete it. That the book can now be published, and that it appears in its present form, is due largely to his kindness and care, and to his knowledge of the district.

There were two drafts of the book in manuscript, an earlier, incomplete, and not all of it quite recent, and a later, also incomplete, but very carefully written. It was upon the latter that Alfred Harker was working during the last months of his life. It was clearly the final draft, almost ready, as far as it extended, for the press. Apart from certain brief additions, mentioned below, and a few unimportant corrections, the first thirteen chapters of the present volume, and the early paragraphs of

Chapter XIV, are printed from this draft. There (at p. 97, line 10, of this volume) the final draft breaks off. An outline of the whole book, in use during the writing of the final draft, shows clearly the form that the unwritten conclusion of the book would have taken. Chapter XIV would have continued the description of the coast of the mainland from Gairloch to Durness. It would have included excursions from Gairloch to Loch Maree and Poolewe, Loch Ewe with Poolewe and Aultbea, Greenstone Point, Gruinard Bay, and Beinn Ghobhlaich; Ullapool and Loch Broom, the Summer Isles, Coigach, the view of the Assynt Mountains from the sea, Lochinver and Strathan, an excursion from Lochinver to Loch Assynt and Inchnadamph, the Point of Stoer, and Eddrachillis Bay; Loch Cairnbawn, Kylesku, and Glencoul, the islands of Badcall Bay, Scourie, Handa, Loch Laxford, Loch Inchard, Cape Wrath, and Durness. A final chapter, to have been entitled 'The Outer Hebrides', would have included the Shiant Isles, a general account of the Outer Hebrides, Stornoway (with excursions to Callanish and Barvas), Loch Shell, and Loch Seaforth; Tarbert (the route from Skye), Scalpay, Leverburgh, and Rodel; Loch Maddy, in North Uist; Loch Boisdale, in South Uist; Castlebay, in Barra; and finally (though this is queried in the outline) St Kilda.

The earlier draft, which had been laid aside, was briefer and followed a rather different outline; but it contained descriptions of places that would have been described in the sections of the final draft left unwritten, and it has therefore been possible to make use of it in completing the final draft for publication.

As already stated, the final draft breaks off at a point early in Chapter XIV (p. 97, line 10, of this volume). The rest of Chapter XIV and the whole of Chapter XV have been added by Dr Richey, partly from the earlier draft, and partly written by himself. The sections written by Dr Richey are distinguished by inclusion within square brackets. It will be seen that these concluding sections follow in general Alfred Harker's own outline, though no attempt has been made to follow his outline in detail or to include all that he would himself have included. In addition to supplying this conclusion to the book, Dr Richey has made certain

additions to the final draft at earlier points. There were three gaps in
the draft, indicated by blank pages in the manuscript, numbered con-
secutively with the rest and bearing in pencil the subjects of the missing
passages. Thus both the extent and the subjects of what was lacking were
clearly indicated. The first of these passages, on Colonsay (p. 29 f.),
and the third, on the journey from Lochalsh to Loch Carron (p. 63 f.),
have been written by Dr Richey; the second, the latter part of the
paragraph on Loch Duich (p. 62 f.), has been supplied from the earlier
draft. Dr Richey has also added the footnotes on pp. 7, 89, 90. As
in the last two chapters of the book, the sections written by Dr Richey
are included within square brackets. This account will enable the reader
to distinguish precisely, if he wishes to do so, those portions of the text
of the volume which are from Alfred Harker's final draft, those which are
from his earlier and less carefully revised draft, and those which are by
another hand.

Dr Richey has also provided the Table of Contents, the Geological
and Topographical Maps, the Table of Geological Formations, the
Glossary of Geological Terms, Minerals, and Rocks, and the Select List
of Geological Maps of the West Highlands and Hebrides. There
was no evidence as to Alfred Harker's intentions in these respects; but
the addition of maps and a glossary appeared likely to add to the interest
and usefulness of the volume.

As far as could be ascertained, Alfred Harker had not chosen a title
for the book. The title under which it now appears has been chosen to
indicate its character and the purpose he had in view.

Nearly all the figures which illustrate the book had been drawn and
described by Alfred Harker himself, and were ready for reproduction.
With the exception of two (figs. 70 and 71), which he drew from photo-
graphs, they are taken from his own sketches. They represent a small
selection from a long series of outline sketches of the West Highlands
and Isles in twenty-eight small sketch-books, now in the possession of
the Department of Mineralogy and Petrology in the University of
Cambridge. The sketch-books cover the period from 1900 to 1938, and
the great majority of those years are represented. Many of the sketches

were made during the cruises off the west coast which have largely determined the form and itinerary of this book. Two of the figures (figs. 69 and 72) had not been prepared, though their subjects and positions in the text were clearly indicated in the manuscript: these have been supplied from the sketch-books. The descriptions or lettering of a few others required completion.

The photograph of Alfred Harker, prefixed to the biographical sketch contributed by Sir Albert Seward, was taken by Mr H. H. Brindley, Fellow of St John's College, Cambridge. It shows him on board the S.Y. *Killarney*, whilst on one of the cruises off the west coast. It was taken in the Inner Sound, between Raasay and Rona to the west and Applecross in Ross-shire to the east, on 22 June 1936.

J. S. BOYS SMITH

ST JOHN'S COLLEGE
CAMBRIDGE
August 1940

Alfred Harker in 1936

ALFRED HARKER

(1859–1939)

Readers of this posthumous book who did not know Dr Harker may like to have an opportunity of learning something about him as a geologist and as a man: others, already familiar with the author's scientific attainments, may be glad to be reminded of the main facts of his life.

Alfred Harker's life is a record of consistent application to a branch of geological science of which he became a master; he made an international reputation by his original and stimulating work on the study of rocks, particularly those of igneous origin. By his teaching and publications he built up a school of Petrology at Cambridge and prepared the way for the establishment, a few years ago, of an independent University Department under Professor C. E. Tilley, F.R.S., whose appointment was heartily welcomed by his old teacher.

The following biographical facts have been selected primarily in order to convey a general impression of the career of a distinguished man of science, who found in a passionate devotion to his chosen subject the satisfying happiness that comes to those who make other interests subordinate to the main purpose and draw no distinction between the professional and recreative aspects of a branch of natural knowledge.

The biographical sketch is divided into two parts; the first part is a general account of Dr Harker's career; in the second part emphasis is laid on his association with Scotland.

I

Alfred Harker was born at Kingston-upon-Hull on 19 February 1859; he died at Cambridge, where as a bachelor he had lived in St John's College for sixty years, on 28 July 1939, in his eighty-first year. He began his education at the Hull and East Riding College and then went to Clewer House School, Windsor; in 1878 he was admitted to St John's

ALFRED HARKER (1859-1939)

College, Cambridge, as a Sizar and two years later elected Foundation Scholar. In 1882 he graduated as eighth Wrangler in the Mathematical Tripos, and in the Easter term of the same year obtained a first class in Part I of the Natural Sciences Tripos; a year later he was placed in the first class in Part II, taking Physics as his chief subject and Geology as the subsidiary subject. His mathematical and physical training was of great value to him in the special field to which he first seriously devoted himself as a geologist. The more important of his earliest contributions to Geology were on slaty cleavage: it is well known that roofing slates can be split almost indefinitely into extremely thin sheets, and it was on the cause and nature of this tendency to cleave that Harker threw new light. The work on cleavage stimulated an interest in the mineral structure and genesis of rocks, especially the various kinds of crystalline rocks of igneous origin. His writings are by no means confined to the branch of Geology known as Petrology—the science of rock structure and origins—in which he became a recognized leader. He had wide interests and was very far from being a narrow specialist; meticulously accurate and thorough in his researches into rock structure, he never lost sight of such major problems as the consolidation of crystalline rocks from molten magmas and the part played by forces that mould the features of the earth's crust upon the internal structure of rocks. His work was always scholarly and philosophical.

In 1884, on the nomination of Professor T. McKenny Hughes, Harker was appointed University Demonstrator in Geology at Cambridge and assumed responsibility for the training of students in Petrology. From the first he gave elementary and advanced lectures with practical demonstrations, and soon after his appointment set himself the task of building up a representative collection of rock-slices for the Cambridge Museum; this collection was his constant care; it now includes about 40,000 specimens, the majority of which were cut from rocks collected by himself at home and abroad. In 1885, Harker was elected Fellow of St John's and re-elected at regular intervals until 1931, when he became a Life Fellow without emolument. In 1918 he was elected to a Readership in Petrology specially created for him: on his resignation in 1931 he accepted the position of Honorary Curator of the Petrological Museum.

His first book was the Sedgwick Prize Essay on *The Bala Volcanic rocks of Caernarvonshire*, published in 1889. In 1895 he supplemented his teaching by the publication of a text-book, *Petrology for Students, an Introduction to the Study of Rocks under the Microscope*: a seventh edition of this very successful and widely used book was issued by the Cambridge University Press in 1935. His next and probably his greatest book, based on courses of advanced lectures, was *The Natural History of Igneous Rocks*, published in 1909. A book on *Metamorphism, a Study of the Transformations of Rock-masses*, was published in 1932. Metamorphism is a fascinating and difficult subject to which he had given special attention; it deals with changes in mineral structure produced in rocks under the influence of pressure or through contact with super-heated masses of material forced upwards from reservoirs far below the earth's surface. Among his earlier contributions to the subject of metamorphism were two papers published by the Geological Society of London in 1891 and 1893 in collaboration with one of his Cambridge colleagues, Dr J. E. Marr, on the Granite of Shap in Westmorland and its effect upon the surrounding rocks into which it had been intruded.

In 1902 Harker was elected Fellow of the Royal Society, and the last honour he received was the Society's award to him of a Royal medal in 1936. From the Geological Society of London, of which he was the President from 1916 to 1918, he received the Murchison medal in 1907 and in 1922 the highest award in the gift of the Society, the Wollaston medal. He was an Honorary Doctor of Laws of McGill University, Montreal, and the University of Edinburgh.

II

The period of his life most relevant to the present sketch is the decade 1895–1905. In 1895, Sir Archibald Geikie, Director of the Geological Survey, in order to strengthen the personnel of the survey staff on the petrological side, invited Harker, who had already made his mark as a petrologist, to become a part-time officer. The arrangement was, that while continuing to perform his duties at Cambridge during two terms each year he should spend the summer months in surveying and mapping Skye and other islands of the Inner Hebrides. This departure from precedent was an unqualified success and abundantly confirmed the wisdom of the Director's innovation. Harker's geological writings on Skye are by no means confined to descriptions of the igneous rocks or to the official publications of the Geological Survey. Much time was spent in the Cuillin hills, and it is probably true to say that the weeks in camp under the shadow of their jagged peaks and awesome precipices provided some of his most satisfying memories. As the writer of the obituary notice in *The Times* (31 July 1939) said—"his brilliant mapping of the Cuillin hills of Skye, completed in 1901, was a tribute not only to his skill as a field geologist and petrographer but also to his prowess as a mountaineer, and brought to a close the historic controversies of Geikie and Judd on the igneous history of that island."

In 1903 the Geological Society published his first paper on the island of Rum. In 1906 the same Society published the classic, and one may add, iconoclastic paper on the Sgùrr of Eigg, an impressive steep-sided, towering ridge of pitchstone, or volcanic glass, that is one of the most conspicuous of Nature's architectural masterpieces in the Western Isles. The basaltic lava-flows and the pitchstone of Eigg were also described in detail in the well-known and exceptionally able survey Memoir on *The Small Isles of Inverness-shire* (*Rum, Canna, Eigg, Muck, etc.*), published in 1908. That memoir and an earlier one on *The Tertiary Igneous rocks of Skye*, have been described as pioneer studies which have served as models for all subsequent survey work in the volcanic centres of Western Scotland. The conclusions reached by Harker, and now generally accepted, may well be called iconoclastic because they were contrary to the attractive interpretation of the Sgùrr by the Director of the Survey, which had appealed to the imagination of geologists and laymen alike. Harker's thorough examination of the Sgùrr led him definitely to dissent from Geikie's conclusions. The Director's view was that a valley eroded in the basaltic plateau had been filled with a series of lava-flows, including the pitchstone, resting on river-gravels. Harker described the pitchstone as a sheet of volcanic rock intruded into, and cutting across, the older basaltic lavas, and the supposed river-gravels as fragmental accumulations of volcanic origin.

Dr Harker contributed a petrological chapter to the Survey Memoir, published in 1903, on *North Arran, South Bute and the Cumbraes*. This included an intensive study of the basic hybrid rocks which form part of the ring of intrusions surrounding the filled-in central pipe of the famous Arran volcano, and of the pitchstone and granites. It is these granitic rocks that form the highest ground and the grandest scenery of the island.

The years spent as an officer of the Scottish Survey were perhaps the most fruitful in Harker's life. Retirement from the Service did not diminish his enthusiastic interest in Scottish Geology: every year, usually twice and not infrequently three times, he travelled from Cambridge to the north, revisiting the scenes of his labours and exploring fresh fields. In 1929 he joined one of the summer cruises among the islands and into the lochs of the west coast on the S.Y. *Killarney*, which sailed from Liverpool, a relaxation which he repeated almost every year, often taking with him one or more Cambridge friends. In celebration of his eightieth birthday he booked a cabin on the *Killarney* for two of the cruises, but in the early summer illness had already begun seriously to undermine his wonderful vitality and cancellation of the anticipated pleasure was philosophically accepted. It was a delightful experience to accompany Harker on one of the cruises: as he stood on the upper deck of the steamer, from time to time consulting a large-scale map, he would occasionally point out to a fellow-passenger, who showed more than a passing interest in the rapidly changing scene, some features of special interest. Despite his modesty and apparent aloofness, he became almost a public character; when the Captain was attacked by eager inquirers he would refer them to Dr Harker who never failed to respond.

Before giving his heart to Scotland, Harker had published several papers on the geology of North Wales, the English Lake District, his native county of Yorkshire, and other districts, but it was the geological

ALFRED HARKER (1859-1939)

riddles of the West Highlands and the Hebrides, over which many battles had been fought by earlier workers, that made the strongest appeal to his keen scientific mind and exercised to the full his highly developed critical faculty. The castles on headlands and by the water's edge were for him much more than picturesque ruins; they spoke to him of men's passions and men's courage. His work in the field and in the laboratory added substantially to the body of facts from which it has been possible to reconstruct the series of events in north-western Europe subsequent to the elevation of the chalk from the sea-floor and continued through several million years. The geological history of the Inner Hebrides is a thrilling story, the story of prolonged and stupendous volcanic activity recorded in the terraced lava-flows on the hill-slopes of Mull, Eigg, and many other islands, and in the pillared walls of Fingal's Cave. The dark crystalline rocks of the Cuillin hills of Skye, the mountains of Rum and other islands that were once buried deep below the surface of the earth and laid bare by the ceaseless operation of eroding and denuding forces; the innumerable dykes that represent the latest phase of the age of fire; the occasional beds of sedimentary material with remains of a rich flora consisting mainly of trees and shrubs closely allied to species still living in the Far East, which in the quiet intervals punctuating the volcanic outbursts were able to colonize the lava-fields of the far-flung Tertiary continent that in the course of ages was converted into the present Scottish Archipelago; the inconceivably old Lewisian rocks of the Outer Hebrides and the North-West Highlands; the rocky hummocks, smoothed and scored in a much later phase of earth history when the western edge of Europe on land and in the sea was overridden by sheets of ice—all these scenic contrasts were to him pictures in the story-book of the earth from which he was able to visualize the process of evolution of the physical world from days when the earth was young.

Like everything he wrote, the book now published, in homage to the memory of a great geologist and a man whom those who knew him well thought of as great in character, is simple and severely restrained, its style reflecting the almost excessive reserve of the writer. As one who had the good fortune to call Harker friend for more than fifty years,

xxii

ALFRED HARKER (1859-1939)

I know that his desire was to encourage others to read the secrets of the Western Isles, to see in their various forms not merely beauty of line, but to experience the joy that comes from ability to follow their building, and the dismemberment of a vast plateau of which they were parts in an age that knew not man.

A whole-hearted lover of Nature, a student whose profession was also his recreation, Harker left a legacy of work well done, and in the hearts of his friends an enduring memory, a sincere esteem, deep affection and a sense of gratitude.

ALBERT C. SEWARD

I. THE ROCKS OF THE WEST HIGHLANDS AND ISLES

The rocks of the West Highlands, including the coastal tract and the western isles, belong in the main to the two extremes of the geological time-scale. On the one hand are the oldest British formations, sedimentary and igneous, both affected in greater or less degree by metamorphism. On the other hand are Tertiary rocks, almost exclusively igneous. Palaeozoic and Mesozoic formations play a much less prominent part. On the mainland we see chiefly the older rocks, and the same is even more true of the long chain of the Outer Hebrides, sometimes named collectively the 'Long Island'. The Tertiary rocks, and in a less degree the Mesozoic, figure mainly on the inner islands. This distribution is connected with the fact that the latter tract has undergone a relative subsidence at a late geological epoch.

The Highlands as a whole are sharply defined by the 'Highland Border Fault', which runs obliquely across Scotland from near Stonehaven to the Clyde, bringing the ancient rocks against the Old Red Sandstone of the Lowlands. Another important dividing line, also due to a master fault, is that of the 'Great Glen', along Loch Ness and Loch Linnhe. In the West Highlands we have to note also an even more remarkable line of discontinuity, traceable from North Sutherland to Skye. This is the 'Moine Overthrust', where older rocks have been driven westward to rest on younger for a distance of many miles.

The Archaean rocks fall into at least three series of more or less metamorphosed sediments (to be called collectively crystalline schists) together with a great system of plutonic rocks (the Lewisian gneisses) and other igneous intrusions. The true succession and mutual relations of the several groups are not yet matters of common agreement, but we shall be content to adopt the view most generally received.

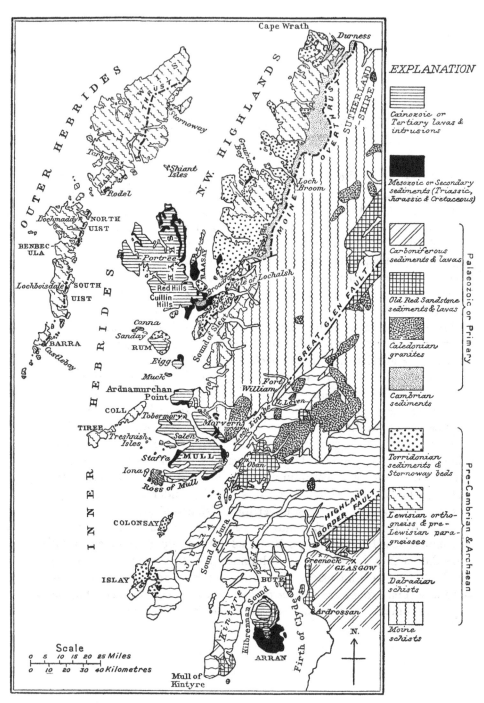

EXPLANATION

Cainozoic or
Tertiary lavas &
intrusions

Mesozoic or Secondary
sediments (Triassic,
Jurassic & Cretaceous)

Carboniferous
sediments & lavas

Old Red Sandstone
sediments & lavas

Caledonian
granites

Cambrian
sediments

Palaeozoic or Primary

Torridonian
sediments &
Stornoway beds

Lewisian ortho-
gneiss & pre-
Lewisian para-
gneisses

Dalradian
schists

Moine
schists

Pre-Cambrian & Archaean

Scale
0 5 10 15 20 25 Miles
0   10   20   30   40 Kilometres

N.

Map I. Geological Sketch-Map of the West Highlands and Islands

TABLE OF GEOLOGICAL FORMATIONS

Era	System	Series and Rock-types	Distribution
Post-Tertiary: Period of 1 million years	Recent and Pleistocene	Records of the Ice Age, raised beaches, river deposits, etc.	Widespread
Unconformity			
Tertiary or Cainozoic: Period of 50 million years	Eocene (Early Tertiary)	Terrestrial lavas, with some lake deposits, and abundant igneous intrusions	Mainly N. half of Inner Hebrides and Arran
Unconformity			
Secondary or Mesozoic: Period of 150 million years	Upper Cretaceous	Meagrely represented by marine deposits (glauconitic greensand and chalk) and white desert sandstone	
		Unconformity	
	Jurassic	Marine and some estuarine deposits, including: 5. Oxfordian shales (marine) 4. Great Estuarine Series, shales and limestones 3. Inferior Oolite (marine), limestone and sandstone 2. Lias (marine), limestone, shales and sandstone 1. Rhaetic (marine)	Mainly N. half of Inner Hebrides and Arran
	Triassic	Terrestrial sediments: bright red sandstone, marl and conglomerate	
Unconformity			
Upper Palaeozoic	Carboniferous	Marine limestones with coal and estuarine deposits, and lavas, including: Coal Measures, Millstone Grit, Carboniferous Limestone and Calciferous Sandstone	
	Old Red Sandstone (O.R.S.)	Terrestrial sediments and lavas in two series separated by an unconformity: 2. Upper; bright red sandstones and lavas Unconformity 1. Lower: dark red sandstones and conglomerates, lavas and intrusions	Mainly around Oban and in the Firth of Clyde and Kintyre areas
Unconformity			
Lower Palaeozoic	Silurian	Great mountain-building crustal movements—the Caledonian orogeny—accompanied by plutonic intrusions	Mainly E. from Moine Overthrust
Period of Palaeozoic 300 million years	Cambrian	Marine deposits including: 4. Limestone and dolomite 3. Serpulite Grit 2. Fucoid Beds (shales) 1. Quartzite	Mainly in Skye and N.W. Highlands
Unconformity			
Pre-Cambrian	Torridonian	Terrestrial red sandstones and conglomerates, with shales	Chiefly in N.W. Highlands and Skye
		Unconformable on Lewisian	
Exact position not known	Dalradian	Altered sediments, including slates, phyllites, quartzite, and limestone, with 'green beds' and lavas	Loch Linnhe to Kintyre and Arran
	Moine	Altered sediments, mainly psammitic and pelitic schists	From Mull north-eastwards, E. of Moine Overthrust
Archaean	Lewisian	Intrusive igneous rocks, mainly gneisses, ultrabasic to acid in composition	N.W. Highlands, Outer Hebrides and parts of Inner Hebrides
Period of Pre-Cambrian and Archaean greater than 1300 million years	Pre-Lewisian	Highly altered sediments, including calcareous rocks, graphite-schists, etc.	Sparse. In N.W. Highlands and Outer and Inner Hebrides

1-2

Pre-Lewisian. The sedimentary formations which most indubitably belong here are for the most part intimately associated with the gneisses, and are in a high grade of metamorphism. They include both argillaceous and arenaceous members (pelitic and psammitic in the common usage) besides pure and impure calcareous rocks.

Lewisian. This formation builds extensive tracts of the North-West Highlands, almost the whole of the Outer Hebrides, and certain of the inner islands. It is a 'complex' of gneisses made by a succession of plutonic intrusions—ultrabasic, basic, intermediate, and acid, in order as named. The earlier members have often suffered mineralogical changes, such as the conversion of pyroxene to hornblende. Further, basic rocks have been attacked by subsequent acid intrusions, giving rise to hybrid products of intermediate composition. Originally the gneisses show only a rough banding, but later events have often set up more pronounced parallel structures. The latest members of the complex are coarse pegmatites.

The Lewisian gneisses have been intersected at some later (but pre-Torridonian) epoch by basic and ultrabasic dykes with a N.W.–S.E. bearing, and these dykes have often suffered a metamorphism of low degree.

Moine. Rocks assigned to this series occupy large areas in the West Highlands, especially between the Great Glen and the Moine Over-thrust. They are sediments in a moderate to high grade of metamorphism. Usually pelitic and psammitic types alternate in considerable thickness of each. The former are chiefly mica-schists, often garnetiferous, and the latter more or less felspathic granulites. Calcareous or partly calcareous rocks are much less in evidence.

Dalradian. This series includes sediments of very various types, and in every grade of metamorphism. The least altered—ordinary slates and schistose grits—are found nearest to the Highland Border. These are the youngest members: dips directed to N.W. might suggest the reverse order, but are due to inversion by subsequent disturbance. There are also important dislocations of the nature of 'slides', which break the succession as now exposed. With advancing metamorphism the common argillaceous rocks become mica-schists, in which distinctive minerals are developed: first biotite (in addition to muscovite), then red garnet, and finally such minerals as staurolite and cyanite (in large crystals) and sillimanite (in tufts of fine needles). The grits are changed to quartzites or to the type styled granulite (containing a notable amount of felspar). The limestones become marbles, or, if less pure, develop such minerals as zoisite, tremolite, and diopside. A very distinctive type of sediment is that of the 'Green Beds', originally rich in chlorite, with some calcite, and becoming hornblende-schists when metamorphosed (often with biotite also).

The rocks known as 'epidiorites' are an important group of basic igneous rocks intruded as sills among the Moine and Dalradian strata and metamorphosed in common with them. These rocks too are in a low grade, chloritic and partly calcareous, while increasing metamorphism gives rise to epidote, plagioclase felspars, abundant hornblende, and often garnet. The 'amphibolites' of this group are distinguished from the corresponding types of the Green Beds by richness in felspar and usually absence of biotite.

Torridonian. This easily recognized formation makes a large spread in the North-West Highlands, where it is seen resting with strong unconformity upon the Lewisian gneisses. Being later than the main metamorphism of the region, it has preserved much of its original characters. It is essentially an arenaceous formation. In the lower part are flaggy or shaly beds, succeeded by grey grits, but the chief thickness is made up of red felspathic sandstone or arkose, often coarse and with pebbly bands. The total thickness is probably not less than 8000 ft.

Cambrian. As the Torridonian was separated from the older Archaean formation by a long interval of time marked by profound erosion, so another such interval divided it from the succeeding Cambrian strata. These consist in their lower 500 or 600 ft. of white quartzites. Next come two minor members, the so-called 'Fucoid Beds', mainly of dolomitic shales, and the 'Serpulite Grit' a dolomitic sandstone weathering in carious fashion. Of small total thickness, these beds are of interest as containing the earliest distinct fossil remains, including the trilobite *Olenellus*. There follow about 1000 ft. of limestones, usually dolomitized and enclosing cherts.

No deposits of Silurian age are found in the West Highlands, but to some epoch during this time must be assigned the gigantic mechanical displacements which have greatly complicated the geological map of this region. They affect a belt of country extending from the north coast of Sutherland as far at least as to the isle of Rum, leaving thus a relatively undisturbed tract to the west, and bringing on the Moine schists abruptly at the eastern margin. The country is cut up into slices by the principal overthrusts, each of which is of the nature of a great reversed fault along a surface inclining eastward at a relatively low angle. The westward displacement of the rocks above such a discontinuity may amount to many miles, and the most usual effect is to cause older rocks to rest on younger. Between two such major surfaces of movement the rocks have sometimes been broken by a closely set series of small reversed faults of steeper grade, causing the same beds to be repeated many times ('imbricate structure'). The major overthrusts themselves are not strictly parallel, so that an earlier one may be cut off by a later, and the whole series is cut off by the most easterly and greatest displacement, the Moine Overthrust. The rocks bordering an important overthrust have often been intensely broken down and sheared, with consequent mineralogical changes, which, however, do not show more than a low grade of metamorphism.

These disturbances are often spoken of as the *Caledonian* system of crust-movements. The same name is applied to an extensive group of plutonic intrusions, accompanied by dykes, which came somewhat later, presumably at a late Silurian epoch. They occur in many parts of the Highlands and Southern Uplands, and it is probable that certain of the granites in the West Highlands belong to this group, but it is difficult to separate them from the rather younger intrusions to be noticed presently.

Old Red Sandstone. Strata of this age are found in Arran and Bute. There is a coarse conglomerate, succeeded by a great thickness of red sandstones with intercalations of conglomerate. Volcanic rocks are interbedded at two horizons, one in the Lower and the other in the Upper

6

Old Red Sandstone. A more important development of volcanic rocks (mainly andesites and basalts) constitutes the Lorne district of Argyllshire. The lavas here are underlain by sandstones with a thick basal conglomerate, resting on the crystalline schists of the Dalradian. Later than the lavas but related to them are the plutonic intrusions, mainly of quartz-diorites and granites, which build Ben Cruachan, Ben Nevis and other prominent mountains. Associated with them are dykes of porphyrite, quartz-porphyry, and lamprophyres with a general N.E.–S.W. direction.

Carboniferous. Rocks of this age are found in Arran and the neighbouring lands, but attain no great importance. Most prominent are a group of basalt lavas and several fossiliferous limestones.

Triassic. The red sandstones and breccias, the New Red Sandstones, which make most of southern Arran are probably of Triassic age rather than Permian.[1] They rest unconformably upon the Carboniferous. Undoubted Triassic rocks, conglomerates and sometimes sandstones, occur at the base of the Mesozoic succession in various parts of the West Highlands, and in places there are scanty representatives of the overlying Rhaetic.

Jurassic. Strata of this age are met with at numerous places in Mull, Skye, etc., but most often they are exposed only in sea-cliffs, where they emerge from beneath the Tertiary basalts. The Lias and Inferior Oolite are both well represented and richly fossiliferous. The Inferior Oolite is succeeded by the Great Estuarine Series, consisting of laminated shales with some intercalations of fresh-water limestones, but marine conditions returned with the Middle Oolite (Oxfordian) division.

Cretaceous. Scanty representatives of Upper Cretaceous strata are found in a few localities. They comprise glauconitic sands, white sandstones, and some relics of chalk.

[1] [The recent identification of basalt lavas of Permian aspect almost at the base of the New Red Sandstone of Arran has added weight to the arguments in favour of placing the lower part of the New Red sedimentary succession with the Permian. It is thought that Dr Harker would have regarded this interpretation with some favour, since no one was more alive than he to the reality of a widespread Permian igneous episode in Scotland.]

I. THE ROCKS OF THE WEST HIGHLANDS AND ISLES

Tertiary. This was an age of vigorous igneous activity throughout the whole region. Most widespread was the outpouring of basaltic lavas. The extensive tracts of basalt now seen in Mull, Morvern, Skye, and elsewhere are probably relics of one great lava-field, which included also a large area in Antrim. Following upon a time of prolonged erosion, the lavas rest on different members of the Mesozoic and older formations. For the most part they issued from very numerous fissures, but there were also eruptions from great central volcanoes, and these may comprise also lavas of other than basaltic composition. At these centres too there are often volcanic agglomerates underlying or interbedded with the lavas, indicating early eruptions of the explosive type. The outcropping of numerous successive flows of basalt on a hillside gives a characteristic terraced appearance. Sometimes there are bands of red clay, where one flow has had time to suffer weathering before being covered by the next flow, and in places there are thin fresh-water deposits with remains of vegetation. The basalts are often amygdaloidal, especially at the surface of a flow, the vesicles being occupied by calcite or chalcedony or most often minerals of the zeolite group.

Breaking through the volcanic rocks at the special centres already established are plutonic intrusions, which may comprise ultrabasic, basic and acid rocks, intruded in that order. In certain cases, however, notably in Mull, the sequence of events is more complex. In addition there are minor intrusions, both concordant and transgressive. Sills of dolerite and basalt are numerous, both in the well-bedded Mesozoic formations below and in the basalt succession. As regards the latter there has been some debate, and for our purposes there will be no need to separate the basalts and their associated sills. Sills of acid rocks are also found, and some of composite nature. Finally must be mentioned basic dykes, usually with a N.W.–S.E. bearing. These are in places extraordinarily numerous, and they have an extension much wider than that of the remaining areas of volcanic rocks.

Post-Tertiary. The latter part of Tertiary time was marked in the West Highlands by faulting, with relative displacement of the faulted blocks, and by profound and long-continued erosion. One effect of this

8

is that Tertiary plutonic rocks, formed under deep-seated conditions, now stand out as mountains up to 3000 ft. in height. In this way the present surface-relief of the country was already determined in its main features before the advent of the Glacial Epoch. Ice-erosion, however, has effected certain modifications, excavating lake-basins, deepening the heads of the fiords, and enlarging the valleys in some mountain-districts. At the culmination of the main glaciation the whole region was overridden by ice from the mainland. This is shown by the glacial striations directed westward (or north-westward farther north) and by the profusion of far-travelled boulders. At the same time sheets of boulder-clay, the ground-moraine of the ice-sheet, were deposited over much of the lower ground. Only certain salient mountain groups (as in Mull and Skye) could withstand the invasion by nourishing each its own local ice-cap. Here and in some other places local glaciers persisted after the withdrawal of the ice-sheet, giving rise to moraines and screes and sometimes to a second set of striae more in conformity with the form of the ground.

Following closely upon the Ice Age there was a general elevation of the land surface. Evidence of this is seen in the raised beaches which fringe much of the coast-line in sheltered situations. The first and highest is 100 ft. or more above sea-level. Other marked stages in the rise are indicated at 40–50 and 20–25 ft. The recent beaches, river-alluvia, and like effects call for no special remark.

II. FROM THE CLYDE TO LOCH FYNE

This route, by the Kyles of Bute, is a very popular approach to the West Highlands. For some 15 miles below Glasgow we have Lower Carboniferous rocks on both sides, the most noticeable being the basalt lavas which build the Kilpatrick Hills to the north and make also a long stretch of the left bank. At Dumbarton, where the 'Rock', made by a plug of dolerite, is a conspicuous landmark, the Carboniferous gives place on the right bank to Old Red Sandstone. Passing Greenock, we catch a glimpse of the Highland mountains, the first being Ben Lomond and the summits about the head of Loch Long, including the 'Cobbler'. The mouth of the Gare Loch opens in front. The Highland Border Fault, throwing the Old Red Sandstone conglomerate against the Dalradian Series, comes to the coast just W. of Helensburgh, and cuts off Roseneath Point. In a trip up Gare Loch one may make a first acquaintance with the youngest and least altered members of the Dalradian. Crossing the fault, we may see the schistose grits at Rhu, phyllites at Roseneath, and mica-schists in the upper part of the loch. All this, as well as upper Loch Long, is easily accessible also by rail.

Fig. 1. View from Ashton (Gourock)

Q, Hunter's Quay; *HL*, Holy Loch; *L*, Loch Long; *K*, Kilcreggan; *G*, Gare Loch; *H*, Helensburgh; *F*, approximate position of the Highland Border Fault; *g*, schistose grits; *p*, phyllites; *m*, mica-schists; *o*, Old Red Sandstone (mainly conglomerate).

At Gourock (a more convenient starting-point than Glasgow) we see the opening of Loch Long and Holy Loch, and obtain a general view of this part of the Highland border (fig. 1). Loch Long shows the same succession as Gare Loch; but the advancing metamorphism as we pass into the interior of the region is made clearer by the more crystalline appearance of the schists in the upper reach of the loch and the appearance of dark mica as well as light.

The next port of call is Dunoon, which is the natural centre for the eastern part of the Cowal district. The town itself stands on the higher raised beach, but the coast southward presents a good section of some of the younger members of the Dalradian succession. First are seen the 'Dunoon Phyllites', belonging to a slate-belt which can be traced from Bute to the Tay at Dunkeld. They are fine sericite-schists of various colours, the green variety being chloritic. The strong schistosity can be seen to cut across a system of acute folds on a small scale. There are also bands of schistose grit and limestone. About a mile S. of Dunoon the phyllites give place to a belt of schistose pebbly grit, the pebbles largely of felspar as well as quartz. This is well seen at the Bull Rock and in a large quarry which is a conspicuous feature viewed from a passing steamer. Both phyllites and grits are traversed by chlorite-quartz and albite-quartz veins. Farther S., and extending as far as Innellan, is a group of alternating phyllites and pebbly grits. The N.W. dip of the whole series being due to inversion, this group is the youngest member seen. The beds structurally below the Dunoon Phyllites may be seen on the S. side of Holy Loch. Here again phyllites are closely associated with more siliceous rocks and with bands of pebbly grit. An epidiorite sill a little beyond the pier at Hunter's Quay contains hornblende with chlorite, epidote and albite.

At Innellan we meet again the Border Fault which, running S.W., cuts off a triangular area at Toward Point composed of Old Red Sandstone. A little S. of Innellan pier is a badly crushed mass of serpentine. The neighbouring grits have been metamorphosed, with the production of biotite. Several N.W.–S.E. basalt dykes of Tertiary age are seen on the coast N. of the pier. Rounding Toward Point and approaching Rothesay (p. 15), we see the granite mountains of Arran appearing over the low ground of Bute. The steamer now makes for the Kyles of Bute.

The East Kyle is mostly grassed or wooded on its shores, but the rocks are well seen at the narrows, in the hills above Caladh Castle W. of the entrance to Loch Riddon, and along this picturesque little loch which has a large gravel flat at the head. The low-grade schists with only sericitic mica and chlorite have now given place to mica-schists with biotite, and in these little crystals of albite begin to be noticeable. These rocks may be examined from Colintraive. By following the road northward from here one may visit the Glendaruel valley, beyond the head of Loch Riddon which is in a broad belt of the Green Beds (p. 5). They are at first epidote-biotite-albite-schists, but hornblende comes in as we pass northward. Some half-mile N.W. of Glendaruel House, on the N.W. edge of the Green Beds, is a band of limestone, the same that is known elsewhere as the Loch Tay Limestone.

The West Kyle shows bolder features than the East. From Tighnabruaich one may take the road to Otter Ferry on the shore of Loch Fyne. At Melldalloch and Auchalick Bay to the W. are seen the same Green Beds and limestone as at Glendaruel; and for half a mile on the coast N. of the bay the mica-schists contain garnet, marking a rise in the grade of metamorphism.

Emerging from the Kyle, we see the low W. coast of Bute with the flat island Inchmarnock and the Kintyre coast as far as Skipness. This latter is made of mica-schists, the continuation of those of Cowal. The next call is at East Loch Tarbet, on Loch Fyne, and on the coast just N. of this is seen the same succession as at Auchalick Bay, the garnetiferous mica-schists coming in at North Bay. Beyond this schistose grits occupy the coast for 6 miles. From near Brenfield Bay, and all along the shores of Loch Gilp, is exposed the member of the Dalradian series known as the Ardrishaig Phyllites, which can be examined from Ardrishaig. They are soft phyllitic rocks, usually more or less calcareous and often containing thin seams of limestone. Intruded among them are numerous sills of the epidiorite group (p. 5), composed mainly of hornblende, epidote, and albite. The same rocks are seen on both sides of upper Loch Fyne. On the N.W. side there are numerous intrusions of quartz-porphyry, one large mass being quarried at Furnace. Inverary makes a convenient centre. Here too is a ferry to St Catherine's, where the epidiorite ('blue-stone') has been largely quarried in the past. Farther down the loch-side, S.W. of Strachur, the phyllite group gives place to garnetiferous mica-schists.

III. FROM THE CLYDE TO THE MULL OF KINTYRE

(a) BY THE FIRTH OF CLYDE

The Cloch lighthouse, opposite Dunoon, may be taken as our starting-point. Turning southward we have a view of Great Cumbrae and southern Bute, with the mountains of Arran in the distance. The Ayrshire coast as far as Largs, most of Great Cumbrae, and the coast of Bute down to Kilchattan Bay are of Old Red Sandstone. Little Cumbrae, however, is made by basalt lavas of Lower Carboniferous age. The separate flows can be picked out by eye, and show a gentle anti-clinal arrangement. The same rocks are repeated in the southern end of Bute.

Arran now comes more plainly into view, the mountains (of Tertiary granite) rising behind the strip of seaboard, a much-faulted area of Old Red Sandstone with Carboniferous strata (p. 21), while Holy Island stands out prominently to the S. Before Brodick Bay the Triassic sandstones come on, and these make most of southern Arran, though with many intrusions of Tertiary age. Clauchlands Point is due to a strong sill of crinanite (p. 19), and Holy Island is made by a thick sill of orthophyre. Farther on the flat island Pladda (with lighthouse) is also due to a sill in the Trias. Away in the south Ailsa Craig is a conspicuous object. It is built by an intrusion of a type of microgranite containing the soda-amphibole riebeckite. The rock is quarried, and is specially valued as a material for curling-stones. All the salient features of southern Arran are due to sill-intrusions. Kildonan Castle stands on a composite sill; inland Auchenhew and other hills are carved out of thick sills of dolerite; another composite sill makes Bennan Head.

Leaving Arran, we approach the Mull of Kintyre. The eastern part, S. of Campbeltown Loch, is of Old Red Sandstone, capped for a short distance on the S.E. coast by an outlier of Carboniferous lavas. Sanda and Sheep islands are also of Old Red Sandstone. To the W. the Dalradian schists emerge, making the Mull proper and rising to 1400 ft. Rounding the Mull we gain a view of part of the Antrim coast. The most noticeable features are due to Tertiary rocks, viz. Fair Head, made by a dolerite intrusion breaking through Carboniferous strata, and to the W. Rathlin Island, an outlying part of the Antrim basalt plateau.

(b) BY KILBRENNAN SOUND

This route carries us near enough to the southern end of Bute to observe something of its constitution (p. 15). A series of basic lavas, of Carboniferous age and of more than one petrographical type, make the country-rock, but they have been invaded by later sill intrusions, which figure prominently (pp. 16–17).

On the N.E. coast of Arran, which we now approach, the seaward slope is a much broken belt of Old Red Sandstone and Carboniferous rocks; but, faulted against this and rising between it and the foot of the granite mountains, the Dalradian schists make a ridge running from North Glen Sannox north-westward. They are bounded eastward by a prolongation of the Highland Border Fault. The extreme north of the island, at the Cock of Arran, is made by a patch of Triassic sandstone, easily detected. The 'Cock' itself, which can be picked out by a field-glass, is a great fallen block of sandstone resting on the beach. Along the coast immediately W. of this, a place of numerous small landslips and stone-falls, fallen blocks lie thickly.

The Dalradian schists now come to the coast, and presently we are looking into Loch Ranza (fig. 6, p. 21), with a glimpse of Caisteal Abhail, second only to Goat Fell in height and shapeliness among the Arran mountains. The schists make a broad strip along the west coast of Arran as far as Dougrie, a distance of 12 miles. It is part of a ring which nearly surrounds the circular area of granite, from which the schists dip outward, so that their strike is generally parallel to the coast. There is an almost continuous fringe of raised beaches, at various levels. The rather monotonous stretch of the Kintyre peninsula, which is in sight from Skipness Point southward, is likewise made by low-grade Dalradian schists.

Farther S. there is more variety, especially the coming in of crystalline limestone, always full of epidiorite sills. Alternations of these two occupy an area N. of Campbeltown Loch (fig. 2), but the low point itself is made by a small patch of Carboniferous basalts, brought down by a fault. A conspicuous feature is Davarr Island, composed of trachyte, apparently a small plug-like intrusion of Carboniferous age. South of the loch the seaboard is made by a tract of Old Red Sandstone; but behind this the Dalradian schists emerge to make the higher ground. The dip of the Old Red Sandstone changes gradually from E. to S., and before reaching Southend an outlier of Lower Carboniferous lavas comes on, capping the cliffs. Southend, the terminus of the high road, stands on a high raised beach. A little N.W. of the hamlet is a 'neck' of Lower Old Red Sandstone volcanic agglomerate breaking through the conglomerate, and another is exposed on the coast at Keil Point. West of this the Dalradian schists re-emerge to make the Mull of Kintyre (p. 23).

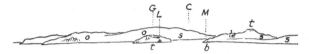

Fig. 2. Approaching Campbeltown

G, Beinn Ghuilean; L, Lighthouse on Davarr Island; C, entrance of Campbeltown Loch; M, Macringan's Point; s, Dalradian schists; le, crystalline limestone alternating with sills of epidiorite; o, Old Red Sandstone; b, basalt and t, trachyte, of Carboniferous age.

IV. BUTE AND ARRAN

(a) THE ISLAND OF BUTE

The Highland Border Fault crosses Bute in a S.S.W. direction from Rothesay along the narrow Loch Fad to Scalpsie Bay on the opposite coast. The larger, north-westerly part consists of Dalradian schists, the continuation of those of Cowal; the rest is of Old Red Sandstone with some Carboniferous. Most of the points of interest are within easy reach of Rothesay. The Dalradian schists are all in a low grade of metamorphism, characterized by sericitic mica and chlorite: only in the extreme N., along the coast of the West Kyle, does biotite begin to appear. The slate-belt (Dunoon Phyllites, p. 10) crosses the island from Kames Bay to Ettrick Bay; but the road between these places runs entirely over a succession of raised beaches, of which the highest makes a considerable spread. There are some belts of schistose grit, one seen at Ardbeg Point, N. of Rothesay, and another along the W. side of Loch Fad. The Old Red Sandstone may be examined along the coast E. of Rothesay to Bogany Point, and farther S. at Ascog is a patch of the Lower Old Red Sandstone lavas. On the road leading S. from Rothesay, near the Church,

Fig. 3. Composite Sill of South Bute, looking S.W.: Arran in the distance

may be seen one of the E.–W. dykes of quartz-dolerite assigned to a Permo-Carboniferous age. A number of Tertiary basalt dykes intersect the Old Red Sandstone, and it is noticeable that they run nearly N.–S., instead of in the usual N.W.–S.E. direction.

The other centre of interest in Bute is Kilchattan, reached by road from Rothesay or directly by sea. A succession of raised beaches occupies the neck of land between Kilchattan Bay and Stravannan Bay, to the W., Kingarth church standing on the highest level. The peninsula thus cut off is made by a series of beds dipping to S.W. The Upper Old Red Sandstone (fringed by a beach) makes a strip for a mile beyond the village, and by following the coast farther we may cross various members of the Lower Carboniferous. Excepting the Calciferous Sandstone at the base, they are all volcanic rocks, including various types of porphyritic basalt and, in one flow, a more felspathic rock allied to bostonite. There are also a number of necks of volcanic agglomerate. Of the intrusions one is of unusual interest. It is first seen W. of Glencallum

Bay, where it forms the headland Roinn Clumhach and makes a prominent escarpment running up on to Tor Mòr. Here it is merely a sill composed of pitchstone. It then breaks away irregularly, and expands so as to cover the hills farther west (fig. 3). Here it is of composite nature: the central and principal member a hypersthene-dolerite, the upper and lower borders a fine-grained dolerite, and injected between central and marginal members a white quartz-porphyry. The junctions are irregular, and show much interaction between basic and acid rocks. This sill is one of the Tertiary system. In places it cuts across an older (Permo-Carboniferous) sill of olivine-dolerite, which is well seen near Garroch Head, where it comes down to the coast on both sides of the headland.

(b) THE CIRCUIT OF ARRAN

Although rocks of widely different ages have their part in the constitution of Arran, the main features are everywhere determined by intrusions of Tertiary igneous rocks. Most prominent is the fine group of granite mountains occupying most of the northern part of the island. The granite covers a nearly circular area of 7 miles diameter, and seems to have been forced upward in dome fashion. The great fault separating Old Red Sandstone from Dalradian, which runs in a nearly straight line across Scotland, is diverted and partly cut out by the granite, the schists now making a nearly complete ring about the intrusion and dipping away from the centre. In central Arran a smaller elliptical area, breaking through Old and New Red Sandstones, consists of a large volcanic vent of Tertiary age bordered by plutonic intrusions. In the southern half of the island, while the country-rock is New Red Sandstone, all the salient features are due to Tertiary sill intrusions.

The circuit of Arran may be by land or by water, for in most parts the road follows the coast. We will begin at Brodick as the most popular resort. The red sandstones with interbedded breccias exposed on the shore belong to the Lower Trias [Permian of Map II]. East of the pier they are cut by numerous basalt dykes, often so much weathered as to cause deep clefts. Following the shore eastward towards Clauchlands Point one may see some well-known sills of spherulitic felsite and

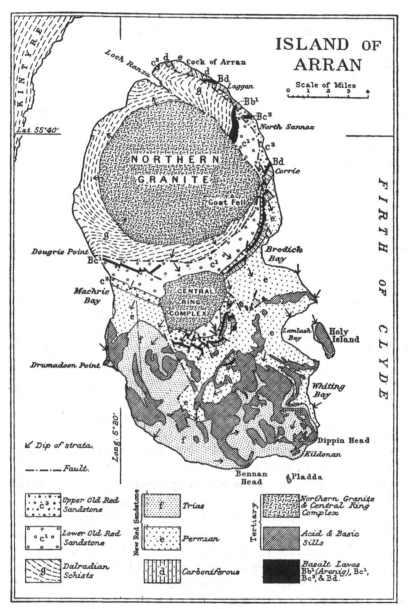

Map II. Geological Sketch-Map of Arran

[Reprinted from Fig. 8, British Regional Geology: Scotland: The Tertiary
Volcanic Districts (*Mem. Geol. Surv.*) 1935. By permission of the Controller
of H.M. Stationery Office.]

pitchstone, and other pitchstones, in the form of dykes, are to be found above in the neighbourhood of South Corrygills Farm. The Clauchland Hills and the Point are made by a strong sill of crinanite, rich in analcime, a type represented elsewhere in Arran. On the shore W. of Brodick pier, two small volcanic vents break through the sandstone, the second and larger one close to the first milestone. They show merely a tumultuous assemblage of blocks of sandstone, representing explosions of gas.

Brodick commands a view of the south-eastern part of the mountain tract (fig. 4). The rock is a biotite-granite. Goat Fell is easy of ascent: note the jointing and deep weathering of the summit crags. A walk up

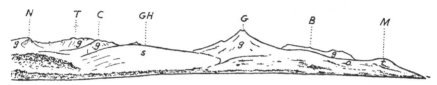

Fig. 4. Seen from near Brodick Pier

N, Beinn Nuis; T, Beinn Tarsuinn; C, Beinn a' Chliabhain; GH, Glenshant Hill; G, Goat Fell; B, Am Binnein; M, Maol Donn; s, Dalradian schists; o, Old Red Sandstone; t, Trias; g, granite.

Glen Rosa is more instructive. After Glenrosa House the track runs for a mile over Old Red Sandstone; then (after crossing the great fault) for three furlongs over Dalradian schists, before reaching the granite. The Dalradian rocks are better exhibited on Glenshant Hill E. of the glen; they are mostly of psammitic types (schistose grits and gritty slates).

An interesting excursion may be made up Glen Cloy S.W. from Brodick. For about 1½ miles the path is on a raised beach, and much of the ground beyond is drift-covered, but the burn shows good sections of the Triassic sandstones with various Tertiary intrusions. At a bend of the stream at Kilmichael a composite sill is well exposed: its interior is of spherulitic felsite and the outer part of andesite (or tholeiite) with a glassy selvage. Continuing up Glen Dubh along a right-hand fork of the stream to the higher tributaries, we reach the plutonic girdle of the great volcanic vent [Central Ring Complex of Map II], consisting of gabbro and granite with hybrid products between the two (fig. 5).

These last-mentioned rocks are more easily reached by the String Road which passes westward from Brodick along the N. margin of the vent-complex. About 3 miles out, soon after crossing the watershed, the road passes from Old Red Sandstone on to granite, and in a quarry at the first burn are good illustrations of gabbro attacked by the granite magma with the production of various intermediate mixtures. Farther on there are exposures of norite on the slopes above. By quitting the road

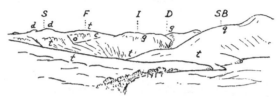

Fig. 5. Looking up Glen Dubh, from below its confluence with Glen Ormidale

S, The Sheeans; F, Creag nam Fitheach; I, Creag na h'Iolaire; D, upper Glen Dubh; SB, Sgiath Bhàn; o, Old Red Sandstone and c, Carboniferous Limestone in faulted strip; t, Trias; g, gabbro invaded by granite; d, hybrid quartz-dolerite intrusions of the Sheeans; q, intrusive quartz-porphyry.

near Glenloig one may reach the agglomerate of the vent. It has a gritty matrix enclosing pebbles and fragments of quartzite, vein-quartz, and various igneous rocks.

The road from Brodick northward runs, as in many other parts of Arran, on the 25-ft. raised beach, overlooked by the old sea-cliff with caves. About Merkland Point and about 2½ miles beyond, the rocks exposed on the shore belong to the lowest member of the Trias, a bright red sandstone with strong current-bedding; but the S.E. dip brings in the Carboniferous as we approach Corrie. First (in descending order) is a group of shales and sandstones assigned to the Coal Measures; but, after passing the Hotel, we come on the Carboniferous Limestone group. There are several distinct limestone beds, of which the most prominent, seen near the old pier, is the 'Corrie Limestone', crowded with *Productus giganteus*. The Calciferous Sandstone group is represented almost wholly by volcanic rocks, olivine-basalt lavas with a coarse agglomerate at the base. These occupy the shore for 350 yards N. of the School. The underlying Old Red Sandstone succeeds. On one of the numerous lines of fault barytes is mined at the foot of Glen Sannox. A walk up this glen affords a view of some of the finest of the granite mountains, Cir Mhòr and Caisteal Abhail.

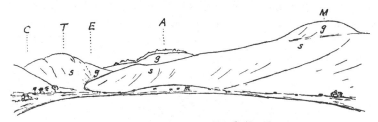

Fig. 6. Looking up Loch Ranza

C, Glen Chalmadale; *T*, Torr Nead an Eoin; *E*, Glen Easan Biorach; *A*, crest of Caisteal Abhail in the distance; *M*, Meall Mòr; *s*, Dalradian schists; *g*, granite; *m*, moraine.

The much-broken strip of country along the N.E. coast is made up of Old Red Sandstone and Carboniferous strata, the Trias reappearing only in the extreme N. about the Cock of Arran (p. 14). The road, however, turns inland up North Glen Sannox, and after about a mile crosses the continuation of the Highland Border Fault. The first of the older rocks encountered is a group of much-altered basic lavas.[1] These are succeeded by schistose grits, in places conglomeratic. At its highest point the road runs for a short distance on the margin of the granite, then descends over rather gritty schists into Glen Chalmadale. Farther on, E. of Loch Ranza, are good slates, which have once been quarried.

Loch Ranza is a convenient centre (fig. 6) from which one may reach the N. coast, or by a hill-track the N.E. coast, or again by Glen Easan

[1] On no grounds that are apparent these rocks have been assigned, if doubtfully, to a Lower Palaeozoic age.

Biorach the granite mountains. From here the road follows the coast closely along almost the whole of the W. side of the island; and for 11 miles it runs over Dalradian schists (though almost always on the 25 ft raised beach) at a distance of $\frac{1}{2}$–1$\frac{1}{2}$ miles from the granite margin. The schists, whether gritty or slaty, are all in a low grade of metamorphism. Sometimes, as at North Thundergay and again about Imachar, they show a fine plication of the 'concertina' type.

At Dougrie the road crosses the great fault (concealed under the 100-ft. beach), passing on to Old Red Sandstone and soon on to the Trias, which is the country-rock of all southern Arran. In it are numerous large sills of quartz-porphyry, and one of these is crossed when the road

Fig. 7. Holy Island and King's Cross Point
t, Trias; *c*, crinanite sill; *o*, thick sill of orthophyre.

turns inland for some distance before coming down to Blackwater Foot. From here one may examine the coast northward. A quartz-porphyry sill, with strong columnar jointing, makes a marked feature at Drumadoon Point. On the shore from 1 to 2 miles N. of this are a number of dykes of andesite or tholeiite and pitchstone, the two types sometimes associated in composite dykes.

About a mile out from Blackwater Foot another large sill of quartz-porphyry is encountered. For a long distance there is little of interest to be observed, though the extraordinary profusion of basalt dykes exposed on the south shore is worthy of note. Bennan Head is made by a composite sill, the quartz-porphyry in this case having split an earlier basic sill, modified remains of which are seen on the two margins. The sill which makes Dippin Head, and is crossed by the road, is of crinanite

22

(p. 19), and a crinanite dyke may be observed just S. of the School at Whiting Bay. From this place one may explore Glenashdale, with acid and composite sills.

King's Cross Point is made by another crinanite sill, which extends also into Holy Island (fig. 7). There it and the Trias into which it is intruded make only the lower ground in the southern half of the island, the main bulk of which, rising to over 1000 ft., is made by a thick sill of orthophyre. It is an alkaline type containing sodic amphibole and pyroxene (riebeckite and aegirine). This place, as well as Clauchlands Point (p. 19), can be reached from Lamlash.

V. FROM THE MULL OF KINTYRE TO OBAN

(a) BY THE SOUND OF JURA

The view of Kintyre still shows an extensive tract of Dalradian rocks, chiefly mica-schists and schistose grits; but an outlying patch of Lower Carboniferous lavas comes to the coast to form the southern horn of Machrihanish Bay. The 3½-mile stretch of blown sand which makes the Machrihanish Golf Course (reached from Campbeltown) is a conspicuous feature. At the close of the Glacial Period the Mull, S. of Campbeltown Loch, was an island, and the present isthmus consists of raised beaches overlain by fresh-water alluvium. Islay and the Paps of Jura are seen in the north, and the summits of the granite mountains of Arran appear over the low ground of Kintyre. The island of Gigha (p. 27) is made by a massive sill of the epidiorite group (p. 5) with some smaller ones, intruded in the Dalradian succession. As we approach the Sound, the white peaks of Jura are seen more distinctly (fig. 8). Almost the whole of this long island is built of quartzite of Dalradian age, though other members make an appearance on the coast (p. 27).

The sea-lochs which form such deep indentations in the mainland are beyond observation from a passing steamer. Note on the map their N.E.–S.W. direction, determined by the regular strike of the Dalradian series. The longest, West Loch Tarbert, dividing Kintyre from Knapdale, is reached from Tarbert, Loch Fyne (p. 26). The rocks about Loch Caolisport, which comes next, are mostly phyllitic mica-schists (correlated with the Ardrishaig Phyllites, p. 12), containing dark as well as light mica, and with many small epidiorite sills. This is a rather remote place, but is accessible from Ardrishaig. Loch Sween, also reached from Ardrishaig, presents more points of note, and Tayvallich is a convenient centre. The rocks in this neighbourhood are pebbly quartzites and sills of epidiorite (calc-chlorite-albite-schist), well seen in a walk across the isthmus to Carsaig Bay on the Sound of Jura. More interesting are the rocks which overlie the quartzites; they may be seen on a cross-

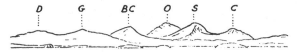

Fig. 8. The Mountains of Jura, from the S.E.

D, Dubh-Bheinn; G, Glas-Bheinn; BC, Beinn Chaolais; O, Beinn an Oir; S, Beinn Siantaidh; C, Corra-Bheinn.

country excursion S.W. of Tayvallich. First, at 1–1½ miles from the village, comes a group of black slates and impure limestones, and these are succeeded by dark green igneous rocks, which make a ridge running from here to the end of the peninsula. They are basic lavas of albitic composition (spilites), often amygdaloidal, and they represent an important volcanic episode in the Dalradian sequence.

Passing the scattered small islands near Craignish Point we look into the Gulf of Corrievreckan, famous for its whirlpool on the N. side, which divides Jura from Scarba. The latter island, a conspicuous landmark, is built of the Jura quartzite, with the Boulder Bed (p. 27) at the base and a strip of black slates fringing the coast. We have passed the wide entrance to Loch Melfort, more conveniently approached from Oban (p. 32). The small islands N. of Scarba are made almost wholly from the Boulder Bed; but Luing, on the other side of the Sound, and some

small islets (including Fladda with its lighthouse) are all of the black slates, known as the Easdale Slates. Abandoned quarries may be seen in many places. The 'Isles of the Sea' (p. 26) are in view to the west, and in the north-west the mountains of Mull. The black slates continue into Seil, and on the smaller island of Easdale (p. 33) are still quarried. Just beyond this the coast of Seil is made by a patch of the Lower Old Red Sandstone andesitic lavas, an outlier of the Lorne plateau (p. 30), and the same rocks make the island of Insh.

The route lies now past the mouth of Loch Feochan, which intersects the Lorne plateau, and by the Sound of Kerrera. The hinterland is still of the andesitic lavas, but the cliffs are made by the underlying conglomerate. On the opposite side the andesites figure on the Kerrera coast for some distance, but the island is built essentially of the black slates covered unconformably by the conglomerate.

(b) BY THE SOUND OF ISLAY

An alternative route to the north passes between Islay and Jura. The S.E. coast of Islay is made by a group of phyllites; but the quartzite emerges to form the high ground behind, and, in accordance with the general N.N.E. strike, comes to the coast as we approach the Sound. These formations are prolonged into Jura, the phyllite group making the southern point of that island. The quartzite thus appears in general on both sides of the Sound, rising in Jura to 2500 ft. in the 'Paps'. On the Islay side, however, some associated members find a place. The Port Askaig Boulder Bed (p. 27) comes on 2 miles before that place is reached and continues beyond it, but quartzite hills are seen again to the north. Numerous basalt dykes intersect the quartzite of the Jura coast and have the usual (Tertiary) direction, N.W.–S.E. Another feature here is the great spread of raised beaches, composed of white quartzite pebbles and often bare of vegetation. They occur at various levels up to 100 ft. and more.

Emerging from the Sound at Rudh' a' Mhàil, and passing the mouth of Loch Tarbert, which cuts deeply into the island of Jura, we see on the left the low islands (nearly united) Colonsay and Oronsay (p. 29), and soon the familiar landmark of Scarba comes out on the right. The route takes us past the Garvellach Isles, known also as the 'Isles of the Sea' (fig. 9). These islands consist of two distinct members of the Dalradian sequence, a white or pale limestone and a group which may be named collectively the Boulder Beds, largely of quartzite. These rocks have been thrown into a succession of sharp folds, and a fieldglass enables one to pick out the white limestone in the cores of the folds. The boulders, large and small, are found on a closer examination to include pieces of the limestone and a variety of igneous rocks like those occurring at Port Askaig and elsewhere.

Fig. 9. Garvellach Isles, from the N.W.

D, Dùn Chonnuill; G, Garbh Eileach; C, A' Chùli; L, Sgeir Leth a' Chuain; N, Eileach an Naoimh: behind are Scarba (S) and part of Jura (J).

In the view of Mull the most conspicuous summits are those of Ben Buie, a gabbro mountain, and Creach Beinn, built up mainly by a complex of 'cone-sheets' (fig. 29, p. 50). The Old Red Sandstone which forms the S.W. end of Kerrera is well seen, dipping to N.W., as we approach the Sound leading to Oban (p. 25).

VI. JURA, ISLAY, AND COLONSAY

This chapter will deal briefly with some of the less frequented of the Western Isles, which nevertheless have regular communications by steamer. Excepting later intrusions and the usual superficial deposits, these islands are built wholly of ancient rocks.

The regular approach to Jura and Islay is by way of Tarbert, Loch Fyne (p. 24). The road across the low isthmus to West Loch Tarbert crosses epidote-chlorite-schists, a prolongation of the belt of Green Beds seen on the W. coast of the Cowal peninsula (p. 11). The west loch is very shallow at the head, and has low wooded shores. On issuing from it the island of Gigha is directly ahead. Though phyllites and quartzite are the country-rocks, the strong features are made by epidiorite sills. They are of the amphibolite type, sometimes carrying garnet.

(a) JURA

Jura is 28 miles long, its shape being determined by the strike of the rocks. These consist essentially of a succession of quartzites dipping to E.S.E.: only on the E. coast higher beds come on, chiefly phyllites, with sills of epidiorite. Most of the island is deer forest, but there is some small population near the landing-place on the S.E. coast. Here is a little bay excavated in the phyllites and guarded by a broken chain o islets (the Small Isles), parts of a sill of epidiorite. Above a green slope the bare quartzite rises steeply to the familiar Paps of Jura. The phyllites come on between this bay and the S. point of the island; but here too the epidiorite makes the prominent features, including the little island of Brosdale.

(b) ISLAY

The geology of Islay shows much more variety. The quartzite and phyllites of Jura are prolonged into the S.E. part of the larger island, but elsewhere other formations enter. The E. coast of Islay and the W. of Jura have already been noticed (p. 25). The place of call for the northern part of the island is Port Askaig. This is the best place to examine the Boulder Bed, already noticed in other localities and well known also on Schichallion in Perthshire. It has been conjectured by some to represent an ancient glacial boulder-clay. The matrix is a biotite-schist, becoming gritty or pebbly and sometimes calcareous. Scattered through this are boulders, large and small, belonging mostly to varieties of granite and quartz-syenite. The place of this formation is between quartzites above and a group of phyllites and limestones below. It is well exposed in the little harbour and beside the Bridgend road, which crosses the island in a S.W. direction. About a mile out it is faulted against a dark blue limestone, which after some two miles is succeeded by a fine phyllite. The phyllite group, with occasional intercalations of limestone, continues nearly to Bridgend. This place, however, and Bowmore are situated on an area of grits assigned to the Torridonian. There is a considerable raised beach on the shore of Loch Indaal. The long straight roads which connect these two places with Port Ellen in the south run over an almost unbroken flat of peat.

Port Ellen has a direct steamer service from West Loch Tarbert, passing S. of Gigha (fig. 10). The village stands on phyllites, while some strong epidiorite sills make the eastern horn of the bay: on the N. and W. is the main quartzite, a continuation of that of Jura.

Fig. 10. *Approaching Port Ellen*

O, Oa peninsula; *T*, island of Texa; *L*, Port Ellen lighthouse; *R*, Rudh'
a' Chuirn; *q*, quartzite; *p*, phyllites with sills of (*e*) epidiorite.

One other place worthy of a visit is Portnahaven, near the extremity of the peninsula called the Rhinns of Islay, reached by road from Bridgend. The western part of Islay beyond Loch Indaal is cut off by an important N.–S. fault, perhaps connected with the Great Glen Fault. Of the area thus separated the northern part consists of Torridonian grits, but the southern part, comprising the Rhinns, is of Lewisian gneiss. The rocks are mainly acid gneisses which have suffered greatly from shearing, resulting in granulitization. They are intersected by intrusions of basic rocks, dykes and larger masses, with a general N.E.–S.W. direction. Of these some retain much of their original characters (diorites, etc.) while others have received a schistose structure (amphibolites). These basic intrusions are well shown on the coast between Portnahaven and Rhinns Point and on the island of Orsay, where the lighthouse stands.

(c) COLONSAY

[The low-lying islands of Colonsay and its smaller neighbour Oronsay, connected at half-tide, are composed almost entirely of Torridonian strata, from under which Lewisian gneiss emerges in the extreme N.E. Viewed from the sea they appear rocky and barren, but the interior is most fertile. Scalasaig, the port of call, is on the E. coast of Colonsay, and from there several roads diverge, with a southerly branch leading to Oronsay.

The Torridonian strata are very varied and consist of grits, epidotic grits, sandstones, flags, phyllites and, near the top of the succession, a considerable limestone. A general easterly dip of the beds towards Scalasaig brings on higher horizons in this direction, and the limestone is well exposed as a dark grey sandy banded rock at the monument S.E. of the village. It outcrops again towards the N. end of the island on the S. shore of Kiloran Bay, where the structure is synclinal.

The Torridonian strata are much puckered and were affected at a later date by cleavage. The crustal movements recorded occurred in two stages, separated in time by the introduction of Caledonian intrusions including diorites and many lamprophyres. Some lamprophyres, however, were later than the second stage of movement.

The pier at Scalasdale is built on diorite, a black and white variety, which forms part of an intrusive mass exposed for some three-quarters of a mile. There are other small related intrusions of kentallenite, hornblendite and syenite in the island. An interesting occurrence of hornblendite at Kiloran Bay encloses quartzite boulders marginally replaced by quartz-syenite as a result of chemical reaction between the magma and the quartzite. There are also several sheets and dykes of lamprophyre, some of which cut the plutonic rocks, although probably of the same general Caledonian date. Among later dykes, some probably Permian, others Tertiary, the most notable example is an ouachitite with large phenocrysts of hornblende, biotite and augite ranging up to about $1\frac{1}{2}$ in. in diameter. The dyke extends along the hills N.E. of Lower Kilchattan.

29

A topographical feature occurring at intervals in both islands is a pre-Glacial raised-beach platform at a level of about 135 ft. O.D. The platform backed by a cliff is well developed around the promontory W. of Kiloran Bay. The post-Glacial raised beaches are not conspicuous except at Kilchattan, but caves bordering the 25-ft. beach of the Mesolithic or early Neolithic period are remarkably large and numerous. In the floor of one of these, on the S. side of Kiloran Bay, bones of the ox and other animals have been found as well as a broken bone-needle and other signs of human habitation. From the shell-mound of Caisteal nan Gillean in Oronsay bone harpoon-heads and other implements together with bones of the great auk and of a variety of animals were obtained.]

VII. THE OBAN DISTRICT

Oban is reached by steamer from the south (p. 23); by road from Ardrishaig, passing Loch Melfort, etc. (p. 32); or by rail via Loch Awe and the Pass of Brander (p. 34). The town stands on the edge of the wide spread of Lower Old Red Sandstone lavas known as the Lorne plateau, the underlying basal conglomerate of the Old Red Sandstone emerging on the coast. This rests unconformably upon the much older Easdale Slates (p. 25), a member of the Dalradian succession. The same rocks appear on the island of Kerrera, which closes in the bay, while beyond are seen some of the mountains of Mull (fig. 11).

The black slates are exposed in the harbour and on the foreshore of the Esplanade, the conglomerate forming a bold escarpment above. At the War Memorial the conglomerate, with intercalated beds of sandstone, comes down to the shore, dipping inland (E.S.E.). Note here a large erratic of the Ben Cruachan granite. The pebbles and boulders in the conglomerate are of andesitic lavas like those of the plateau, showing that volcanic activity had already begun in some neighbouring area before the deposition of the conglomerate. Near here we obtain a good view down the Sound of Kerrera (fig. 12). The black slates are seen again between here and Dunollie Castle, and the unconformity may be examined on the shore. The road runs on a raised beach-level, a rather broad wave-cut platform, and the conglomerate escarpment is an old sea-cliff with caves and a prominent stack. The view to the N.W. shows the entrance to the Sound of Mull between Mull and Morvern (fig. 13).

Fig. 11. Looking across Oban Bay from the North Pier

The Mull mountains seen are *M*, Ben More in the distance; *D*, Sgùrr Dearg; *F*, Mannir nam Fiadh; and *G*, Dun da Ghavithe; all composed of Tertiary igneous rocks. In front is the N.E. end of Kerrera with the Hutchison Monument (*H*). The rocks seen on the island are: *s*, black Easdale Slates; *c*, Lower Old Red Sandstone conglomerate; *a*, andesitic lavas; *r*, raised beach.

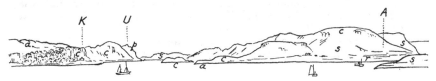

Fig. 12. Looking down the Sound of Kerrera

K, Kilbowie House; *U*, Dun Uabairtich; *A*, Ardantrive Bay; *s*, black Easdale Slates; *c*, Lower Old Red Sandstone conglomerate; *a*, andesitic lavas; *p*, columnar porphyrite intrusion; *r*, raised beach.

Much of the town is built on a raised beach; but there are exposures of lavas and conglomerate, and faulted inliers of the black slate group. The andesitic lavas, mostly somewhat altered, are seen in Glen Cruitten, Glen Shellach, etc. They are mostly of a basic type, and near Oban often contain pseudomorphs after hornblende. An instructive walk is that along the Gallanach Road S.W. of the town. The conglomerate makes the cliffs above, and is seen by the roadside; but the black slates are below, and are conspicuous on the shore at Dungallan House and Kilbowie. They

are well seen too as we approach the Kerrera Ferry, and are there intersected by brown-weathering dykes of minette, of Caledonian or Old Red
Sandstone age. At about 2 miles from Oban, before reaching the little
bay with shipyard, a conspicuous columnar-jointed crag overlooks the
road. It is a plug-like intrusion of a mica-porphyrite or andesite. The
place is named Dun Uabairtich (fig. 12). This walk affords good views
of Kerrera across the Sound. What is seen is mainly the black slates
surmounted on the higher ground by the conglomerates; but a faulted-in
strip of the lavas makes the coast for some $1\frac{1}{2}$ miles or more.

Fig. 13. *Looking towards the Sound of Mull, from Dunollie Road*

In the foreground are the N.E. point of Kerrera, made by black slates (*s*) and
Maiden Island (*M*) of conglomerate (*c*) and a dolerite intrusion. Lismore lighthouse
(*L*) marks the entrance to the Sound. To the left are Duart Castle (*D*) and Dun
da Ghaoithe, made of Tertiary basalt and agglomerate (*b* and *a*). On the right
is An Sleagach (*S*), part of the granite hills of Morvern (*g*), and in front of this
the low island of Lismore, composed of limestone (*l*) of Dalradian age.

Oban is an excellent centre for short excursions, both by land and by
water. A favourite drive is to the Pass of Melfort. After passing a faulted
inlier of the slates near Soroba House the road runs over the main lava
plateau. The rocks here are mainly pyroxene-andesites, but a branch
road on the right leads by Kilbride to Ardentallan, on the N. side of Loch
Feochan, where a group of basaltic lavas comes in. The road crosses the
broad flat at the head of Loch Feochan, with the River Nell, and then
follows for some miles the shore of the shallow loch. The scenery is that
typical of the Lorne plateau—broken hills but showing more or less
clearly the terraced appearance due to a succession of distinct flows. At
Kilninver we turn inland, and in some 5 miles reach the Pass. The
modern road is carried over the hill; but here we are at the base of the
lava succession, and, the conglomerate being absent in this area, Dal-

radian rocks are seen at the beginning of the road leading down to Melfort House. Here they are phyllites, belonging to the Ardrishaig group; lower down are banded mica-schists and near the pier bands of quartzose limestone.

Another excursion is to Easdale. The route is the same, viz. the Ardrishaig road, as far as Kilninver, where our road turns away inland, and in about 1½ miles reaches Loch Seil. Here a fault brings up the black slates, and for the next 1½ miles the road runs on the slates but close to the line of fault. It then takes a northward turn to cross the Sound of Seil where it is narrow enough to be spanned by a bridge. Seil is one of the slate islands, and quarrying operations may be noticed at

Fig. 14. Seen from above Easdale Village

L, Luing; S, Scarba, with Lunga in front; J and I parts of Jura and Islay; D, Eilean Dubh Mòr; G, Garbh Eileach (Isles of the Sea); E, S.E. corner of Easdale Island; t, Torridonian (in W. Islay); s, Easdale Slates; b, Boulder Bed; q, Quartzite.

Balvicar on the left before the road turns away to cross the island. Approaching the Easdale village (Ellanbeich) we have a comprehensive view seaward (fig. 14). The strong escarpment above the road is made by an outlying area of the Lower Old Red Sandstone lavas, which occupies the N.W. part of Seil. Quarrying operations in the village have ceased since the deep pit was filled by an inrush of the sea in the great storm of December 1879 (the same that destroyed the first Tay Bridge); but the quarries on Easdale Island are still working. The slates are often plicated on a small scale, and enclose crystals of pyrites.

Loch Awe may be reached either by road or by rail. Leaving the town by the Connel road we see a good section of the Old Red Sandstone conglomerate with its interbedded sandstones, but the road quickly rises on to the lava plateau. Along the shores of Loch Etive, which is followed for some miles, road-making operations have exposed fresh material. Connel Ferry is the place of the 'Falls of Lora'. With a falling tide the water of the loch, piling up at this narrow place, gives rise to formidable rapids. Something of the same effect is to be observed at Loch Creran and Loch Leven. The fine outline of Ben Cruachan comes into view in front. Nearing Taynuilt we see across the loch the Bonawe quarries (p. 35) worked in the same granite mass that builds Ben Cruachan, and there is a good view up the long upper reach of Loch Etive. About 2½ miles beyond Taynuilt the road crosses the River Awe, and presently enters the Pass of Brander, where there is barely room for river, road, and railway. Ben Cruachan rises steeply above, the granite border being only ½–¾ mile from the road, which is thus well within the aureole of metamorphism. The rocks exposed at the roadside are the semi-calcareous Ardrishaig Phyllites, now in the condition of evidently crystalline mica-schists with various new-formed lime-silicates between the films of mica. Note the very numerous porphyrite dykes with N.E.–S.W. bearing, satellites of the Ben Cruachan intrusion, of Old Red Sandstone age. The road turns N.E. along the shore of the loch, cutting into a group of dark biotite-schists, which a little beyond the Hotel and Station give place to quartz-schists and quartzites. Loch Awe is peculiar in that its inlet (from Glen Orchy) and its outlet (by the R. Awe along the Pass of Brander) are at one end, and the long south-westerly stretch of 23 miles ends in a cul-de-sac. The original outlet was by Ford to Loch Craignish and the Sound of Jura. When this was blocked by the ice-sheet, the rising waters found a new exit at the N. end. The establishment of this was facilitated by the coincidence of the Pass with an important line of fault, along which erosion was specially effective. Hence the evidently newly cut gorge and the rapid river rushing down to Loch Etive.

To reach Loch Awe by Glen Nant we follow the road already described until it crosses the River Nant about ½ mile beyond Taynuilt. Here a branch road, starting steeply up-hill, leads up Glen Nant. Much of the glen is densely wooded, but roadside exposures show still the Lower Old Red Sandstone lavas, including good hypersthene-andesites and at

Taylor's Leap rhyolites. There are also dykes of mica-porphyrite. Where the road rises out of the glen we come to the edge of the lava plateau and pass on to the Dalradian rocks, mostly black slates. Over the tarn on the left there is a good view of the Ben Cruachan range. There are a number of epidiorite sills in the Dalradian series. A massive one is seen at Kilchrenan, where the road ends on the shore of Loch Awe, and the slopes on the opposite side of the loch, above Port Sonachan, are all of epidiorite. These rocks consist of fibrous hornblende, epidote granules, and albite, with some chlorite.

Kerrera may be reached either directly from Oban or by the ferry, 2 miles along the Gallanach Road. Its geology is a repetition of that of the mainland.

Loch Etive is most conveniently visited by steamer, starting from Achnacloich. For the first 2 miles we are still within the boundaries of the Lorne lava plateau, with raised beaches fringing the shores. The black slates then appear on the north side, and in another mile the large quarries of Bonawe show that we have come to the granite. The long reach of 11 miles north-eastward lies within the granite tract. This is of complex constitution. There are two principal intrusions, both of boldly transgressive habit, the younger breaking through the older. The outer ring, which includes Ben Cruachan itself, is of quartz-diorite; the inner mass, which builds Ben Starav and neighbouring summits, is of biotite-granite.

Loch Melfort may be reached by water from Oban, the route being past the slate islands Seil, Easdale, and Luing, and through the narrow passage of Cuan Sound, with its strong tidal race. Other steamer excursions are to the 'Isles of the Sea' (p. 26), Craignure and Loch Spelve (pp. 49–50), Salen, Mull (p. 43), Tobermory (p. 45), Loch Aline (p. 43), Lismore (p. 36), and Loch Creran (p. 36).

VIII. LOCH LINNHE AND LOCH LEVEN

Starting from Oban, we pass Ganavan Sands, with the headland of conglomerate beyond, and have a glimpse of the lower reach of Loch Etive to Connel Bridge with Ben Cruachan in the distance. The peninsula between Ardmucknish Bay and Loch Creran is mostly covered by raised beaches and peat, but the Dalradian schists crop out on the shore, as we round the point to enter Loch Linnhe. The long green island of Lismore is practically all built of limestone; but the small islands, haunt of the grey seal, are mostly of quartzite, as is also Eilean Dubh, higher up the channel. The narrow sinuous passages on either side of the

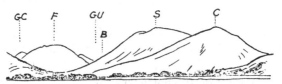

Fig. 15. Looking up Loch Creran to the Narrows

B, railway bridge; *GC*, Glen Creran; *F*, Beinn Fhionnlaidh; *GU*, Glen Ure; *S*, Beinn Sguliaird; *C*, Creach Bheinn. All is of Dalradian rocks, viz. black schists with bands of quartzite, except the summits of the mountains on the right, which are made by the margin of the Ben Cruachan granites with a border of metamorphosed schists.

flat island of Eriska give access to Loch Creran. The rocks surrounding the loch are black schists with quartzose bands ('Leven Schists'), and there is the usual fringe of raised beaches (fig. 15). The headland immediately before Port Appin is of quartzite, and the village stands on a raised beach. In front is the island Shuna, made of the same limestone as Lismore ('Ballachulish Limestone'). On the right Beinn Bheithir (Ben Vair) comes into view, the granite mountain which overlooks Ballachulish on the other side (fig. 16).

Having cleared Lismore, we have an uninterrupted view of the Morvern coast across the loch. The little bay called Loch Corry can be visited from Oban, and the pink biotite-granite of Morvern is well exposed in the ice-worn rocks about its entrance. Not far beyond this the Morvern granite ends, and the hills to the north are made by rocks of the Moine Series. It is to be remembered that the important dividing line of the Great Glen Fault runs along Loch Linnhe, passing on the farther side of Lismore. The next main valley, Glen Tarbert, leading to Loch Sunart, divides Morvern from Ardgour.

The mouth of Loch Leven opens on the right, and calls for a digression. It is best explored from Ballachulish Ferry as a centre. This is situated on the granite tract of Beinn Bheithir and near its eastern border. On the railway, E. of the Station, the schists near the contact are highly metamorphosed with the production of cordierite, andalusite, and sillimanite. The granite is seen along the road westward, and has been quarried. It is more precisely a granodiorite, and is remarkable for the number of enclosed dark patches, representing digested fragments of schist. The western edge of the granite is crossed at Kentallen pier, and metamorphosed schists and limestones are

Fig. 16. *Beinn Bheithir, seen from the S.W.*

The mountain is made by an intrusion of granite, but Dalradian strata make the lower slopes on this side. At *q* the Kentallen quartzite quarry is seen over the flank of Ardsheal Hill.

exposed here and along the road. At the mouth of the little inlet of Kentallen Bay a dark crystalline rock is seen by the roadside, the so-called kentallenite, a basic type related to the granite and of the same Old Red Sandstone age. Farther on a white quartzite is quarried on the hillside. This is the Appin Quartzite, part of a series which builds the Ardsheal peninsula S.W. of Kentallen, but is more easily examined on the N. side of Loch Leven. To see this, cross the ferry, and follow the Fort William road to Onich. First, near the Church, are old quarries in the Ballachulish Slates; then near the Nether Lochaber Monument is seen the Appin Quartzite, often pebbly; farther on, along the shore, the Appin Limestone, showing in places dark alternating with cream-coloured stripes; finally the Appin Phyllite, a sericitic schist, sometimes with biotite. Going on to Onich village, we cross the same succession in reverse order owing to a sharp isoclinal fold; and at the

37

headland all is cut off by an important thrust or 'slide', which brings on flaggy beds (Eilde Flags), probably part of the Moine Series.

The view up Loch Leven from the Ferry includes the mouth of Glen Coe with some of its surrounding mountains (fig. 17). The road runs first upon mica-schists (Leven Schists) until these are cut off by a 'slide'. On both sides of Ballachulish village we may examine the Ballachulish Slates, well-cleaved black slates with crystals of pyrites, and farther on the Ballachulish Limestone.

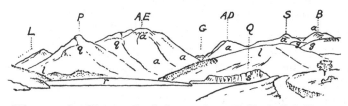

Fig. 17. Looking up Loch Leven from Ballachulish Ferry

L, upper Loch Leven; *P*, Pap of Glencoe; *AE*, Aonach Eagach; *G*, Glen Coe; *AD*, Aonach Dubh; *Q*, Ballachulish slate-quarries; *S*, Stob Coire nan Lochan; *B*, Bidean nam Bian; *q*, Glen Coe Quartzite; *l*, Ballachulish Limestone; *s*, Ballachulish Slates; *a*, andesitic and other lavas within the cauldron subsidence; *g*, granite of the fault-intrusion; *r*, raised beach. The cauldron subsidence just includes the summits of Aonach Eagach and Bidean nam Bian.

Glen Coe must not be passed without a brief notice. The alluvial flat in the lower part of the glen overlies the limestone, which is exposed again near the Clachaig Inn. Here the less pure parts are represented by some interesting rock-types—actinolite-schists and zoisite-biotite-schists with crystals of zoisite sometimes measuring as much as $1\frac{1}{2}$ in. in length. The upper part of Glen Coe lies within a remarkable 'cauldron subsidence'. A sunken area of elliptical shape and 9 by 5 miles in extent is enclosed by a continuous curved fault, which must have a throw of some thousands of feet. It is occupied mainly by an outlier of the Lower Old Red Sandstone lavas. Its southern boundary is cut by a part of the Ben Cruachan granite mass, and a broken ring of granitic and allied intrusions follows the fault on its outer side. The road crosses the fault at the nearer end of

Loch Achtriochtan, and runs for a mile over the Leven Schists, on which the volcanic rocks unconformably lie. At Achtriochtan House we come upon andesitic lavas, and a mile beyond this is the locality of the red manganiferous epidote, withamite, which occupies veins and druses in a hornblende-andesite. A little farther on the road crosses rhyolites, and other types of lavas may be seen on the neighbouring hillsides. These volcanic rocks are intersected by very numerous N.E.–S.W. dykes, mostly of porphyrite. They are related to the Ben Cruachan granite centre, and correspond with the swarm seen in the Pass of Brander on the other side of the granite (p. 34).

Fig. 18. Part of Ardgour, seen from the Fort William road

C, the Corran; *E*, Sgùrr na h-Eanchainne and *L*, Druim Leathad nam Fias. Between these two mountains is Glen Scaddle and in the distance Sgùrr Dhomhnuill (*D*), the highest point of Ardgour, its upper part made by a gneissic granite; *G*, Glen Cona.

Returning now to Loch Linnhe, we come to the narrows made by the Corran, a spit of gravel attributed to a fluvio-glacial origin. From here the Ardgour district can be explored, but most visitors are content to pass it by. It may be viewed from the steamer or from the Fort William road. It is a monotonous tract of Moine sediments, chiefly felspathic and quartzose granulites. Intruded among these, however, is a large mass of basic hornblendic rock, partly foliated, which makes the mountains Sgùrr na h-Eanchainne and Druim Leathad nam Fias, and is intersected by Glen Scaddle (fig. 18). It belongs probably to the epidiorite group, but seems to have the form of a large boss, and is surrounded by an aureole of metamorphism. On the right as we proceed are the higher mountains of Lochaber, culminating in Ben Nevis, made by a granite intrusion.

The coast on this side and the area about Fort William are built of a group of rocks known as the Eilde Flags and correlated with some part of the Moine succession. It is separated by a 'slide' from the formations exposed to the S.E. The rocks are well-bedded, with more and less micaceous layers, varying in character from a siliceous granulite or quartzite to a garnetiferous mica-schist. Examples may be seen in the quarry at Bridge of Nevis and in many other places. An excursion should be made up Glen Nevis. This lies wholly within the aureole of metamorphism of the Ben Nevis granite, and the rocks are modified accordingly. The 'slide' is crossed near Glen Nevis House. Then, for a mile, comes the Ballachulish Limestone, its more impure parts largely replaced by such minerals as tremolite and diopside; after this the Leven Schists, which become converted to a hornfels with cordierite and andalusite. This lower reach of Glen Nevis illustrates the U-shaped cross-section and straight course due to widening by ice-erosion, and there are a number of small 'hanging valleys' on the lower slopes of Ben Nevis. The mountain to the S., Mullach nan Coirean, is made by a separate granite intrusion, which extends down into the glen at the Lower Falls of Nevis at Achriabhach. It is of a type poor in mica.

The granite intrusion which makes the great mass of Ben Nevis (fig. 19) is, like that of Ben Cruachan (p. 35), a double one, and it presents some remarkable features in addition. There is an outer ring, broken on the S. side, and an inner, both having nearly vertical boundaries; so that they may be conceived as having a roughly cylindrical shape. Within the inner granite, and forming the summit of the mountain, is a cylindrical mass of the country-rocks, viz. the Lower Old Red Sandstone lavas resting on Dalradian schists. This is a sunken tract of very limited dimensions, and must be supposed bounded by an elliptical fault like that of the cauldron subsidence of Glen Coe (p. 38). Its sinking and the uprise of the granite magma surrounding it were parts of one mechanical process. Following the regular tourist track from Achintee Farm, we see first lime-silicate-rocks belonging to the Ballachulish Limestone before meeting the outer ring of hornblende-biotite-granite at a point opposite to Glen Nevis House. The first 400 yards of this has

a grey colour, then becoming pink with porphyritic crystals of orthoclase. It is cut by dykes of lamprophyre and porphyrite, but none of
these enters the inner granite. The boundary between outer and inner
is met at the zigzags S. of the lochan and W. of the halfway hut, and is
crossed three times by the path, which then traverses the inner ring of
pink biotite-granite up to 3000 ft. altitude. Above this come the volcanic
rocks, viz. hornblende-andesite and agglomerate, metamorphosed by the
granite. The underlying schists can be seen only in two places, one of

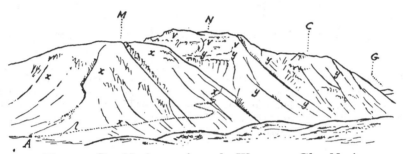

Fig. 19. Ben Nevis seen from the W., across Glen Nevis

M, Meall an t' Suidhe; *N*, Ben Nevis; *C*, Carn Dearg; *G*, upper Glen Nevis;
A, Achintee; *l*, lime-silicate-rocks; *v*, volcanic rocks capping the mountain;
x, outer granite; *y*, inner granite.

which is a little to the right of the track. On this upper part of the
mountain the effects of frost in shattering the surface rocks is very
apparent.

A short excursion from Fort William is a walk along Loch Eil. The
Moine sediments here have a decidedly crystalline aspect, and the more
so as we go westward. About Locheilside and from there to Glenfinnan
are banded biotite-plagioclase-gneisses, the bedding picked out by seams
rich in mica. These rocks are often crumpled and shot through with
pegmatite: some contain garnet.

IX. THROUGH THE SOUND OF MULL

Coming from Oban, we gradually open the view of the Sound as we near the point of Lismore (fig. 20). If the approach be from the S., there are glimpses on the way of Loch Spelve and Loch Don (pp. 49–50). There is a fine panorama of the mainland mountains, made by several granite intrusions of Old Red Sandstone (Caledonian) age, from Ben Cruachan

Fig. 20. Approaching the Sound of Mull from Oban

The Mull mountains seen are S, Sgùrr Dearg; F, Mannir nam Fiadh; G, Dun da Ghaoithe: their upper parts are of Tertiary volcanic agglomerates, the lower of basalt lavas resting on Mesozoic strata. D, Duart Castle; J, Java Point; L, Lismore lighthouse. On the right is the low island of Lismore, made of a Dalradian limestone. The Morvern hills behind (*MM*) are of granite, but the low point Rudh' an Ridire (*R*) is of Moine siliceous granulites. The Morvern coast beyond this is of basalt.

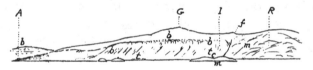

Fig. 21. Part of the Morvern coast, seen from near Craignure

A, Ardtornish Castle; G, Glas Bheinn; I, Inninmore Bay; R, Rudh' an Ridire; m, Moine granulites; c, Carboniferous (Coal-Measure) sandstone in Inninmore Bay; t, Trias sandstone surmounted by Lias; b, Tertiary basalt lavas brought on by the Inninmore fault (*f*).

in the E. to Ben Nevis, seen up Loch Linnhe. After passing Duart Point and Craignure Bay, we have on the right the headland Rudh' an Ridire. This, with the adjacent small islands, is of Moine granulites; but just beyond this, in Inninmore Bay, a fault brings on the Tertiary basalts, which make the cliffs from here onward (fig. 21). As on the Mull side, the basalts belong to the lowest part of the succession. Liassic and

42

Triassic are seen below, and for a short distance in Inninmore Bay Carboniferous sediments; but after Ardtornish Bay, as the result of another fault, these older rocks are below sea-level. A little farther we pass the narrow mouth of Loch Aline. On the shores of the loch the Mesozoic strata reappear, and can be conveniently examined. In addition to Triassic and Liassic there is here one of the few good exposures of Cretaceous strata, consisting of greensand overlain by white sandstone.

The next port of call is Salen. Approaching it, we look into Glen Forsa with its U-shaped cross-section, and the shapely outline of Beinn Talaidh. Farther to the south are seen the sharp summit of Beinn Fhada and the round back of Beinn a' Ghràig with Ben More in the distance.

Fig. 22. Beinn a' Ghràig, seen from the fork of the roads about $2\frac{3}{4}$ miles from Salen

b, basalt; *g*, granite; *f*, felsite of ring-dyke.

Salen, situated in the narrow waist of the island, is a point of vantage for exploring the neighbourhood. The immediate country is made by the basalt lavas with minor intrusions; but the structure of the higher ground is complicated by intrusions of granite. This is well seen by following the road which, rising not more than 100 ft., crosses to the opposite coast of the island. The mountain Beinn a' Ghràig is composed partly of basalt, partly of granite (fig. 22). The junctions are in most places vertical, and a remarkable feature is a narrow 'screen' of basalt caught between two intrusions of granite. This is continuous with a cake of metamorphosed basalt forming the summit, a relic of the 'roof' of the main granite mass. For a closer examination of the rocks, follow the road as far as the sharp bend and take the track leading to Loch Bà. The granite, as seen in an old quarry near Benmore Lodge, is an augite-

43

bearing type, in places containing fayalite. At Sròn a' Chrann-lithe a remarkable 'ring-dyke' may be examined where it comes down to the loch in a dark scar. The glacier which once occupied the valley has left its traces in the moraines on either side and in the glacial gravels at the foot of Loch Bà. There is a raised beach at the head of Loch na Keal, behind which the river Bà is diverted northward. The road along the sea-loch has various points of interest, including views of Ben More (fig. 23). The mountain can be ascended, starting from Dishig, but is more accessible on its southern side (p. 55).

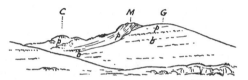

Fig. 23. Ben More seen from the N.N.W.

C, A' Chioch; M, Ben More; G, An Gearna; b, basalt lavas; p, the 'pale group' of lavas, which includes a group of mugearites.

From Salen a visit may be made to Glen Forsa. Its glacial history is in part like that of Loch Bà, and there is a similar diversion of the river behind a raised beach. The glen corresponds roughly with the line along which the Scottish ice-sheet was held up on encountering the local ice-cap, and is marked by a belt of hummocky morainic drift. A remarkable feature of Tertiary igneous activity in Mull is seen in the systems of parallel inclined sheets (cone-sheets), usually of dolerite, dipping towards a common centre, and often so numerous and closely set as to exceed in bulk the country-rocks. Beinn Talaidh (fig. 24) is perhaps the most striking illustration. For a closer appreciation it is worth making a traverse up the Gaodhail River, which debouches some 3 miles up the glen. As exposed there the inclined sheets outweigh the country-rock in the ratio 2 : 1.

The complex of plutonic intrusions in south-eastern Mull is surrounded by a belt of country in which the basalt lavas are somewhat altered, the olivine in particular being destroyed. Such is the nature of the rocks about Salen; but a little way out on the Tobermory road one passes beyond this belt on to fresh olivine-basalts. On the top of the hill Braigh a' Choire Mhòir, ½ mile W. of the village, is a small neck of volcanic agglomerate, breaking through the basalts and pierced in turn by an intrusion of albite-trachyte. Half-way along the coast to Tobermory a similar trachyte occurs at An Sean Chaisteal.

The N.W. division of Mull has less variety of interest, since there are no plutonic intrusions to break the monotony of the basalt plateaux. At Tobermory the base of the volcanic series is exposed at the mouth of the

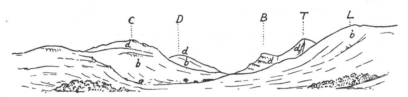

Fig. 24. Looking up Glen Forsa from near its mouth

C, Beinn Chreagach Mhòr; D, Sgùrr Dearg; B, Beinn Bheag; T, Beinn Talaidh; L, Beinn nan Lus; b, basalt lavas; d, parallel inclined sheets (cone-sheets) of various types of dolerite, so closely set as to constitute the main feature of the local geology.

river, near the Distillery: it rests on Triassic sandstone. Higher up the river is a glassy basalt dyke with selvages of tachylyte. The walk along the face of the cliff to Rudha nan Gall is a good place to observe the highly amygdaloidal basalts, rotted by the volcanic gases. Various zeolite minerals occur in the vesicles.

Opposite Tobermory are Auliston Point and the mouth of Loch Sunart, with Ben Hiant a prominent object farther W. Loch Sunart is worth a digression (fig. 25). Auliston Point still shows the well-bedded basalts, and an outlying patch makes Ardslignish on the other side; but entering the loch we soon see the Moine schists emerging on the right and left, well exposed on the coasts and islands. Passing Glenborrodale Castle, we have a glimpse into the narrow mouth of Loch Teacuis; and in the next reach, approaching Salen of Ardnamurchan, Ben Resipol is a conspicuous object in front. This is built of Moine schists, as are also the mountains of Moidart seen to the N.

Loch Sunart divides Morvern from Ardnamurchan, a district presenting more points of interest. The western peninsula is occupied mainly by a Tertiary plutonic complex, with a concentric ring-like disposition of the several types, accompanied by a great system of inclined sheets

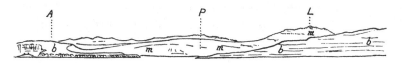

Fig. 25. Entrance to Loch Sunart

A, Ardslignish; P, Auliston Point; L, Ben Laga; m, Moine schists;
b, basalt lavas.

Fig. 26. Ben Hiant, seen across the Sound

M, Maclean's Nose; m, Moine schists (mica-schists, often garnetiferous);
v, volcanic agglomerate of vents; q, quartz-dolerite.

dipping inward (cone-sheets). The earliest incident of the igneous sequence was, however, the opening of a number of volcanic vents, filled with agglomerate. Two of these occur on the eastern and southern flanks of Ben Hiant, one of them running out into the promontory named Maclean's Nose (fig. 26). They are partly cut out by a large intrusive body of quartz-dolerite, which makes the bulk of the mountain. Between here and Kilchoan Bay the country-rock is mainly of Moine schists, but inclined sheets, dipping W.S.W., make up a large part of the ground. These too are chiefly of quartz-dolerite. On the E. side of Kilchoan Bay is a patch of the basalt lavas; on the W., Lower Lias shales and Inferior Oolite limestones with many inclined sheets, now dipping N.N.W. in accordance with the general concentric scheme.

From Kilchoan Ben Hiant is well seen (fig. 27), and may be visited. The high-road is on Moine rocks, of psammitic type and in a very low grade of metamorphism, with the usual inclined sheets. At the sharp turn of the road $1\frac{1}{2}$ miles from Kilchoan take the track eastward, crossing the burn. This traverses a curious medley of rocks, representing the heterogeneous contents of a volcanic vent and offshoots from the quartz-dolerite of the Ben Hiant mass. Besides amygdaloidal basalts there are

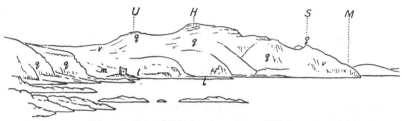

Fig. 27. Ben Hiant, seen from Kilchoan

U, Beinn na h-Urchrach, with Mingary Castle in front; *H*, Ben Hiant; *S*, Stallachan Dubha, with an outlier of quartz-dolerite capping agglomerate; *M*, Maclean's Nose; *m*, Moine schists; *l*, Lower Lias; *v*, volcanic agglomerate; *q*, quartz-dolerite.

patches of Trias conglomerate and Lias limestone. Before the path rejoins the high-road, the Moines reappear; but by striking off southward one may examine the volcanic vent on the E. flank of the mountain. With basaltic and trachytic agglomerate, it includes also considerable patches of lavas, both basalts and augite-trachytes, and in two places there are sheets (perhaps flows) of pitchstone.

Kilchoan is the centre for an examination of the remarkable plutonic complex of Ardnamurchan, but of this our space does not allow more than a very brief notice. The concentric rings, in places irregular or inconstant, consist principally of various types of eucrite and gabbro. A junction between types may sometimes be found by careful search; but

47

it is clear that the several intrusions followed one another very closely. The central and youngest intrusion is of acid nature, and has given rise to hybrid products with the surrounding basic rocks. A rough general view may be gained by following the road northward from Kilchoan to Achnaha. Beginning on metamorphosed basalt lavas, it passes on to typical eucrite near the Free Church and Manse. Some 400 yards beyond this a small quarry by the roadside shows a quartz-felsite dyke cutting the eucrite and enclosing partly-digested fragments of it. At the second sharp angle of the road, in an old gravel-pit, the eucrite becomes

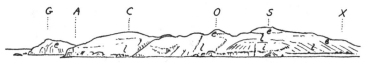

Fig. 28. The coast of Ardnamurchan beyond Kilchoan, looking N. and N.W.

X, the crofts of Ormsaig; *S*, Beinn na Seilg with Maol Buidhe in front and the headland Sròn Bheag; *O*, Beinn nan Ord; *C*, Beinn nan Codhan; *A*, An Acarseid (anchorage); *G*, Garbhlach Mhòr (the Point of Ardnamurchan concealed behind); *l*, Lower Lias shales with inclined sheets dipping inland; *i*, Inferior Oolite sandstone, faulted down to make Sròn Bheag, with inclined sheets; *e*, eucrite making all the higher ground.

so highly felspathic as to pass into an olivine-anorthite-rock (allivalite). After crossing the first burn, the road traverses a belt of quartz-gabbro, passing on to olivine-eucrite again a little before the second burn. This continues for a mile or more, and is succeeded by gabbros, some with biotite. Where the road turns more north-westerly towards Achnaha is very coarse and crumbling eucrite, passing into evidently mixed or hybrid rocks bordering the central granitic area. To see this, with its fringe of hybrid products, a traverse may be made N. and N.E. across country to Glendrian Farm.

The cliffs W. of Kilchoan show Mesozoic rocks, dipping seaward, with inclined sheets, dipping inland, but these disappear before reaching Ardnamurchan Point, where only eucrite is seen (fig. 28).

X. THE CIRCUIT OF MULL

Loch na Keal and Salen Bay, with the intervening strip of low ground, divide Mull into two unequal parts. The smaller, north-western, area is built essentially of Tertiary basalt lavas, with Mesozoic rocks emerging at places on the coast. In the larger, south-eastern area, excluding the Ross peninsula, the same rocks figure, but the dominant factor in the topography is made by a complex sequence of plutonic intrusions, breaking through the volcanic rocks. The geology is far too complicated to be set forth in any brief summary. The concentric disposition seen at some other centres, notably in Ardnamurchan (p. 47), is here carried to further developments. Characteristic features are the arcuate plan of many of the intrusions (ring-dykes of the Geological Survey) and the multiplicity of intrusive sheets inclined inwards (cone-sheets). These latter are not all of basic composition. They were injected at more than one epoch, and there was a certain shifting of the centre to which they are related. This peculiar arrangement of intrusions of many kinds is bound up with a system of concentric anticlinal and synclinal folds, which are clearly displayed in the coastal belt. Their effect is even discernible on the map in the semi-circular sweep of the coast-line and in such details as the shape of Loch Spelve.

The nearest place to observe these structures is Craignure, easily reached from Oban. The bay is hollowed out of a sharp anticline, striking N.W.–S.E. On the shore near the village the Trias conglomerate forms the core of the fold, and is flanked by Lias shales. There is some crushing and faulting, and the rocks are cut by inclined sheets consisting of a felsitic rock (craignurite). More striking is the Loch Don anticline, a few miles to the S. and easily reached from Craignure. Leave the highroad near the little Church at Lochdonhead, and make for the tarn named Loch a' Ghleannain. This lies on the axis of the fold, which strikes due N.–S. and has a core of Dalradian rocks (phyllites with some limestone). This is flanked by Old Red Sandstone basic lavas, dipping steeply off on each side. The Mesozoic strata come on in their turn and then the Tertiary basalts, all dipping sharply away from the axis.

To the traveller from Oban, taking a southerly route,[1] the low ground about the shallow Loch Don is not noticeable, and the first object of interest is the mouth of Loch Spelve (fig. 29). The N.E.–S.W. stretch of the loch corresponds roughly with still another anticline, bringing up the Trias. The mountains behind are of granophyre but riddled in their upper parts by inclined sheets. Croggan, which has occasional steamer connections, is the port of call for Loch Buie, in addition to the normal passenger route by Craignure. The road runs first over basalt lavas, but at Barachandroman, near the end of the loch, is a good exposure of one of the bedded volcanic breccias. The fragments are of basalt and quartzite with some of Moine rocks. The N. shore of Loch Uisg is occupied by an intrusion of granophyre.

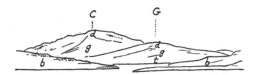

Fig. 29. The mouth of Loch Spelve, looking W.

C, Creach Beinn; G, Glas Bheinn; t, Trias sandstones; b, basalt lavas; g, granophyre; d, mainly inclined sheets of dolerite.

The coast from Loch Spelve to Loch Buie follows an anticlinal axis situated a little off-shore, which brings up the Lias into the lower part of the cliffs in many places. This is well seen at Frank Lockwood's Island and the approach to Loch Buie. Note too the great profusion of N.W. dykes, the 'swarm' related to the Mull plutonic centre. Passing the mouth of Loch Buie, we have a view of Ben Buie. Its nearer slope consists of the agglomerate of an early volcanic vent, cut by many inclined sheets, now dipping northward. This has been partly enveloped by a large intrusion of gabbro or more precisely eucrite, which makes the bulk of the mountain.

---

[1] The popular 'Staffa and Iona' excursion starts north-about and south-about on alternate days.

Coming to Carsaig Bay, observe the island Gamhnach Mhòr, consisting of alkali-syenite, a unique occurrence in this district. The cliffs on the E. side of the bay are much broken by landslips, but on both sides the Mesozoic rocks are seen making the lower part of the cliffs (fig. 30).

Carsaig, which is most conveniently reached from Kinloch (p. 55), has more than one point of interest for the geologist. In the steep burn, with a succession of waterfalls, above Carsaig House is a good section of the Upper Cretaceous strata of Mull. They rest with strong unconformity upon the white Liassic sandstone, and consist of about 40 ft. of greensand with *Exogyra* and other fossils. With little indication of a break, there follows a Tertiary sandstone, best seen in the burn above Feorlin Cottage.

Fig. 30. The coast from Carsaig Bay to Malcolm's Point

M, Malcolm's Point; N, Nun's Pass; C, Carsaig Bay; l, Lias; b, basalt lavas; d, strong sill of dolerite.

The grains show that thoroughly rounded shape which is suggestive of desert conditions at the beginning of Tertiary times in this region. The white sandstone in the cliffs of Carsaig Bay belongs to the Middle Lias. It is remarkable as enclosing large spherical concretions, which weather in relief. There are old quarries in this rock at Nun's Pass, a mile to the S.W. Near this place the low promontory of Rudh' a' Chromain is made by a composite stratiform intrusion, having as its middle member a felsitic rock, flanked on each side by a more basic type (tholeiite). The upper member is remarkable for the number of foreign inclusions which it contains. The most interesting are those which represent some argillaceous sediment in a highly metamorphosed state, showing crystals of anorthite, spinel, and sapphire. Farther on the Mesozoic rocks sink below sea-level, and we can observe some of the stratified deposits which occur in the lowest part of the volcanic succession. About 1½ miles beyond Nun's Pass the section in the cliff shows some 30 ft. of amygdaloidal basalt, followed by 15 ft. of a coarse conglomerate of rolled flint pebbles. This passes up into a bedded basaltic tuff, covered by a dolerite sill. Followed to Malcolm's Point, these stratified deposits thin away, the pebbles disappearing, and near here, at Carsaig Arches, there are only a few feet of sandy tuff.

Three miles W. of Malcolm's Point the basalt lavas are abruptly cut off. An important fault, ranging W.N.W. from here to Bunessan, brings on the Moine Series; first psammitic types (granulites) and then, after another fault, pelitic schists in a high grade of metamorphism. Farther W. comes a large intrusion of granite of Caledonian age, running out into the headland of Ardalanish and forming all the western part of the Ross peninsula. It makes also the Torran rocks, scattered through the sea for 5 or 6 miles to S. and W. It is seen at close quarters as the steamer threads her way between the rocky islets. It is a red biotite-granite, but encloses dark patches of an earlier diorite, and the largest of the little islands is composed of this diorite. The island Soa, seen to the W. as we turn northward for the Sound of Iona, is, however, an outlying fragment of the Lewisian of Iona. Entering the Sound, we can see the old Iona marble quarries on the S.E. coast. The small islands which fringe the coast farther on belong to the Ross of Mull granite, which just fails to reach Iona.

The greater part of Iona consists of Lewisian gneisses with some enclosed patches of pre-Lewisian sediments; but along the eastern coast is a strip, ¼–½ mile wide, of Torridonian rocks. A conglomerate rests upon the Lewisian to the west, and is succeeded on the coast by a group of flagstones, well exhibited near the pier and village. They are cut by a few dykes of camptonite and teschenite. Farther S. an underlying group of shales or slates comes in. It reaches the coast about a mile from the village, and is seen to be metamorphosed by the granite. The conglomerate is seen at the hill beyond, Druim Dhughaill: here and in other places the rocks are much crushed. Some 500 yards farther, in a gully running steeply down to the sea, is the locality of the Iona marble, a strip enveloped by the gneiss. It is a serpentinized forsterite-marble with a patchy arrangement, and associated with it are rocks with radiating needles of tremolite. Other types of metamorphosed sediments occur, in particular actinolite-magnetite-schists. Distinct from all these is a fine-grained 'white rock', which at the quarries makes a narrow strip, but traced northward spreads out to a width of $\frac{1}{4}$ mile. It is a crushed and altered albite-rock, part of the Lewisian complex. The gneiss, as seen here and elsewhere, is mostly of an acid type, especially hornblende-albite-gneiss, but there are bands of coarse dark hornblende-schist and veins of pegmatite. Farther west, at Port na Curaich, the eastern arm of St Columba's Bay, is another patch of marble enclosed in the gneiss. It is much sheared, and encloses nodules or lenticles of hornblende.

A section of the Torridonian of Iona may be taken along the road which crosses the island. Its base is reached about 200 yards beyond the cross-roads. The gneiss is seen by the roadside, and the 'white rock', near its termination, is exposed in the crofts S. of the road. West of the crofts is the Machair, an old raised beach with a sea-cliff. Following the W. coast northward, one may see sills of quartz-porphyry, satellites of the Ross of Mull granite. Varieties of gneiss may be noted about Dùn Cùl Bhuirg, and about 300 yards N.N.E. of this hill is another patch of marble, known as the 'silver-stone'. It is a forsterite-marble spangled with flakes of talc.

There is a ferry from Iona to Fionphort, where the granite of the Ross can be seen, with some dykes of porphyrite and kersantite which intersect it. Motor-boat excursions can be made also to Staffa (p. 56).

Iona commands a good prospect of part of Mull (fig. 31), and as we emerge from the Sound the Treshnish Isles and Staffa come into view in the north (fig. 32). Before leaving the Ross of Mull, however, we shall make a digression to Bunessan, overlooking the inlet Loch na Lathaich. This place is situated nearly on the fault which throws down the Tertiary volcanic rocks against the Moine schists. On the E. side of the loch are the basalts, on the S. the Moines, and on the W. the granite, excepting a little faulted patch of Moine granulites at the western horn of the loch (fig. 33). The Moine schists of this district are in a very high grade of metamorphism, marked by the coming in of cyanite and sillimanite in addition to garnet. Blue crystals of cyanite, associated in places with tourmaline, may be seen on the S.W. side of Loch Assapol inland and S.E. of Bunessan (the N.E. side coincides with the fault bringing on the basalt). West of Bunessan sillimanite comes in, in the form of crowds of minute needles. In this direction, however, is the large granite intrusion,

Fig. 31. Mull seen from Columba Hotel, Iona; looking E. and N.E.

In the foreground is the ice-moulded granite of the Ross; in the distance the bedded basalt lavas of the coast between Loch Scridain and Loch na Keal. B, Bearraich; C, Creach Bheinn with the summit of Ben More (*M*) just showing beyond; G, Gribun cliffs; K, Inchkenneth (Trias, resting on Moine granulites and mica-schists).

Fig. 32. The Treshnish Isles, seen from the S.

BB, Bac Beag; BM, Bac Mòr (Dutchman's Cap); L, Lunga; F, Fladda; CB, Cairn na Burgh; S, Staffa.

which has its own special aureole of metamorphism. Here is developed new sillimanite in stouter crystals, associated with andalusite and cordierite, while garnet is destroyed. These effects are best seen in masses of Moine rocks actually enveloped in the margin of the granite: to examine them follow the road westward for a mile, and strike off S.W. from Bendoran Cottage.

In the basalt area special interest attaches to the 'Ardtun leaf-beds', which have furnished the chief evidence for the Tertiary (Eocene) age of the volcanic series. They form part of a sedimentary group intercalated in the lower part of the succession, and are exposed in the cliffs 2 miles N. of Bunessan. Traced laterally the sections vary somewhat, but include in general two or three leaf-beds alternating with sands and gravels. Running along the shore or in the base of the cliff is a conspicuous intrusive sheet, cutting irregularly through the volcanic rocks. It consists of a partly glassy basalt with a selvage of spherulitic tachylyte.

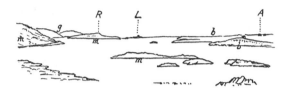

Fig. 33. Seen from Bunessan Hotel, looking N.W.

R, Rudha na Tràighe-maoraich; L, lighthouse; A, Ardtun; m, Moine Series; g, granite of the Ross; b, basalt lavas.

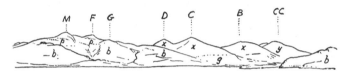

Fig. 34. Seen from Kinloch Hotel, Loch Scridain, looking N. and E.

M, Ben More; F, Beinn Fhada; G, Beinn nan Gobhar; D, Cruachan Dearg; C, Corra Bheinn; B, Beinn a' Mheadhoin; CC, Cruach Choireadail; b, basalt lavas; p, the 'pale group' of lavas; g, granophyre of Derrynaculen; x, gabbro or eucrite with many inclined sheets; y, eucrite-granite hybrid rocks, cut by some inclined sheets.

Bunessan is visited by steamers; but other places on the W. coast of Mull can be reached only by boat or by road. Kinloch, at the head of Loch Scridain, is worth a visit. It commands a comprehensive view including Ben More and some of the mountains on the north side of Glen More (fig. 34). These latter, in the heart of the great plutonic complex, have a highly complicated constitution. Ben More may be ascended from here. The 'pale group' which forms its upper part consists merely of olivine-basalts poorer in iron-oxide than the rest, with an intercalated horizon of the more sodic mugearite type. Carsaig (p. 51) is another locality within easy reach of Kinloch.

Resuming now the regular circuit of Mull, the next objective is the island of Staffa (fig. 35). The jointing of the basalt, seen in many places in western Mull as well as in Canna, Skye, etc., is here developed in a manner which has made the island famous. The effect is due to contraction of the rock on cooling, the columns being perpendicular to the cooling-surface. Where master-fissures have introduced fortuitous surfaces of cooling, the regular parallelism gives place to a more peculiar

Fig. 35. Staffa, seen from the S.W.

C, Boat Cave; F, Fingal's Cave; B, Buchaille: the 'causeway' in front of the cliff affords access to Fingal's Cave from the landing-place near the Buchaille; t, basalt-tuff; x and y, the upper and lower divisions of the thick columnar basalt.

disposition, as is well seen in the stack named the Buchaille and in the Clamshell Cove, a little further north. Less than half of the columns show the ideal hexagonal cross-section. A conspicuous feature is the contrast between the large and regular columns in the lower half of the sheet and the smaller, less regular columns in the upper half. This is due to steady cooling by conduction downward and less steady cooling, complicated by convection, at the upper surface.

Leaving Staffa, we obtain a nearer view of the basaltic Treshnish Isles, backed by the long outline of Tiree and Coll. The well-known Dutchman's Cap and its neighbour Lunga show a terraced appearance which is not due merely to the bedding of the lavas, but primarily to a pre-Glacial wave-cut platform, a feature which can be more or less clearly detected in other parts of the Western Isles (p. 30). It is seen again on Treshnish Point and Caliach Point. Presently Ardnamurchan Point appears ahead, and there remains only the passage of the Sound of Mull, which, in the opposite direction, has been already described (pp. 42–48).

XI. FROM ARDNAMURCHAN POINT TO THE KYLE OF LOCHALSH

Shortly after passing the lighthouse on Ardnamurchan Point we look into Sanna Bay with its stretch of blown sand, the low hills of eucrite rising behind. The N. coast of the peninsula for some 4 miles E. of this is on the fringe of the plutonic complex of Ardnamurchan. Beyond are the mountains of Moidart and Arisaig, part of the great tract of Moine

Fig. 36. *The Small Isles of Inverness-shire, seen from S.S.E.*
MM, Muck; RR, Rum; EE, Eigg; S, part of Skye in the distance; t, Torridonian; o, Great Estuarine Series; b, Tertiary basalts; u, ultrabasic rocks; e, eucrite; f, quartz-felsite; p, pitchstone. Canna and Sanday are hidden behind Rum.

rocks which makes so much of the country N. of the Great Glen. To the N.W. we have a good view of the group of islands known as the Small Isles, with some of the Skye mountains showing in the distance (fig. 36). The steamer passes near enough to enable some of their features to be observed. The Sgùrr of Eigg is conspicuous, carved out of a thick sheet of pitchstone and, in an end-on view, standing out as abruptly as a tower. Below are the basaltic lavas; but towards the N. of the island Jurassic rocks rise gradually above sea-level (fig. 37). As we approach the Sound of Sleat the mainland also claims attention: the white sands of Morar and the depression which marks the place of Loch Morar in its rock-basin, deepest of all the Scottish lochs. Mallaig, accessible by rail or road as well as by sea, is a convenient centre for those who would explore the districts of Morar and Knoydart.

All is of Moine schists, but these present some variety of lithological characters. Pelitic types predominate, garnetiferous mica-schists in particular, about Mallaig and along the road eastwards to Mallaigvaig at the mouth of Loch Nevis. There are also pale bands, only a few inches in thickness, which have been of semi-calcareous composition, and consist now of zoisite and a yellow garnet, with quartz, biotite, and sometimes calcite. Loch Morar is well worth a visit. A common type

Fig. 37. *Part of the E. coast of Eigg*

K, Kildonan; A, An Cruachan; b, basalt; m, a pale band of mugearite in the basalt group; o, Middle Oolite Series, mainly the Great Estuarine Sandstone.

there is a mica-schist containing epidote. Loch Nevis and Loch Hourn can be visited by the motor-boat which carries the mails from Mallaig, and there is a regular ferry to Armadale.

Leaving Mallaig, we have a glimpse up the lower reach of Loch Nevis to Inverie Bay and the surrounding mountains. At Inverie, reached from Mallaig, the rocks are mainly biotite-oligoclase-gneisses or granulites, sometimes with garnet, hornblende, or epidote. On the Skye side of the Sound are Armadale and the village of Ardvasar, the best centre for the Sleat district.

It is to be observed that the great Moine Overthrust, ranging approximately along the Sound of Sleat, cuts off a strip of Skye extending some 10 miles or more from Loch na Dal to near the Point of Sleat. This strip, consisting of Lewisian and Moine rocks, is thus brought against the Torridonian, which makes the higher ground to the N.W. The Moines, as seen near Armadale pier and on the coast about Ardvasar, are mostly mica-schists some containing epidote and others garnet. Similar types may be seen on the coast beyond Tormore, 2 miles to the S.W. The Lewisian gneisses, as displayed along the coast in front of Armadale Castle grounds, are much crushed and sheared as a consequence of the over-thrusting. Across the Sound of Sleat are seen the mountains of

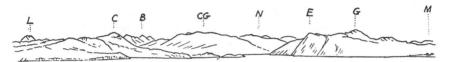

Fig. 38. The mountains of Knoydart, seen from Armadale pier

L, Ladhar Bheinn; C, Sgùrr Coire na Coinnich; B, Meall Buidhe; CG, Sgùrr Coire nan Gobhar; N, mouth of Loch Nevis; E, Sgùrr an Eilein Ghiubhaic; G, Càrn a, Gobhar; M, Mallaigvaig.

Fig. 39. Looking up the Sound of Sleat, from Armadale pier

Y, Sgùrr na Coinnich, and Z, Beinn na Caillich (the mountains overlooking Kyleakin); K, Kyle Rhea; G, Glenelg Bay; C, Beinn a' Chapuill; H, mouth of Loch Hourn; S, Ben Sgriol, with Rudh' Ard Slisneach in front; B, Beinn na Caillich.

Knoydart, culminating in Ladhar Bheinn (Larven), a tract essentially of Moine schists and granulites (fig. 38). The view up the Sound (fig. 39) includes country of much more varied geological constitution. The place of the Moine Overthrust is just E. of Kyle Rhea. On the left are Torridonian mountains; on the right, beyond Loch Hourn, a complicated tract of pre-Lewisian, Lewisian, and Moine.

Isle Ornsay is another possible stopping-place. The Lewisian gneisses, well displayed round the harbour, are mostly of acid types, and have suffered more or less severely from shearing. There are also basic rocks

59

very rich in hornblende. The rock on which the lighthouse stands is an example, of exceptionally coarse texture and carrying abundant red garnet. In the little bay below Duisdale Hotel is a patch of Moine mica-schists, mostly containing epidote. Just beyond this place comes the Moine Overthrust, bringing on the Torridonian rocks. These are not the red felspathic sandstones so widespread farther north but greyish grits with shaly bands.

From Isle Ornsay we look into the mouth of Loch Hourn (a place of much geological interest but not easily visited), and have a nearer view of Ben Sgriol and its neighbours (fig. 40). From here the route along the

Fig. 40. Seen from Isle Ornsay

I, Sandaig island; *B*, Meall Buidhe; *C*, Beinn a' Chapuill; *M*, Beinn Mhialairidh; *S*, Ben Sgriol; *H*, the place of Loch Hourn. The lower hills are built essentially of Lewisian gneisses (*l*), with some narrow infolded strips of Moines (*m*), one of which makes the summit of Beinn Mhialairidh. Moine rocks make the upper part of Beinn a' Chapuill and, owing to a fault, the whole of Ben Sgriol.

Sound of Sleat is near the course of the Moine Overthrust. On the left is a rather monotonous tract of the Torridonian grits; on the right, extending as far as Loch Alsh, a country presenting exceptionally intricate geological relations. Some notion of its structure can be gained by short excursions from Glenelg as a centre.

The rocks are acutely folded on axes having a general N.N.E.–S.S.W. direction. In this direction run numerous strips of pre-Lewisian sediments, in a very high grade of metamorphism, enveloped in the gneiss and involved in the same folding.[1] The Moine Series, deposited upon the Lewisian, has also shared in the folding, and there are infolds of Moine rocks in the area which is mainly of the old gneisses. These

[1] Only the limestones are separated on the Geological Survey map (1-inch Sheet 71).

latter show some variety of composition and structure. Well-banded examples may be examined beside the road leading southward from the pier: they are mostly of a type containing hornblende, biotite, and epidote. Of basic rocks the most interesting are the eclogites, composed essentially of red garnet and green pyroxene. Associated with them are related hornblendic rocks (garnet-amphibolites). They occur in abundance on the low hills E. of Glen Bernera and W. of the path from Glenelg to Ardentoul on Loch Alsh. To see the pre-Lewisian rocks to advantage it is worth making an excursion southwards up Gleann Beag. Calcareous types are seen by the roadside 700 or 800 yards beyond Corrary and on the

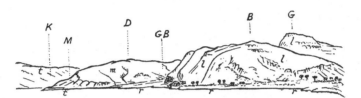

Fig. 41. Glenelg Bay, seen from near the Church, looking N.W.

K, Kyle Rhea; M, Moine Overthrust; D, Druim na Leitire; GB, Glen Bernera; B, Biod Bàn; G, Glas Bheinn; l, Lewisian gneiss; m, Moine schists; t, Torridonian; r, raised beach.

hillside above Baloraid. They are forsterite- and diopside-marbles enclosing nodules of diopside and phlogopite mica. Highly metamorphosed argillaceous rocks may be seen on Druim Iosal, ¾ mile farther on. They are of coarse texture, and contain garnet, staurolite, and cyanite. The Moines may be examined in Glen Bernera, N. of Glenelg, where they are mostly quartzose schists with white mica. The place of the Moine Overthrust at the N. end of the bay is visible from Glenelg, and can be visited by taking the road to the Kyle Rhea ferry (fig. 41). The rocks contiguous to the surface of displacement are sheared to the state of 'mylonites'.

Passing through the narrow Kyle Rhea, with its powerful tidal race, we emerge into Loch Alsh. Directly opposite is displayed another of the great dislocations which have affected the older rocks of the West Highlands. To the W. of Loch Alsh the Torridonian strata have a moderate dip to N.W., but on the N. coast of Loch Alsh they rather suddenly become inverted, and more markedly so as they approach Balmacarra. At the headland E. of the hotel they are cut off by what is known as the Balmacarra Thrust and overridden by the Lewisian gneiss, recognizable by its reddish weathering (fig. 42). As usual, all the rocks bordering the surface of movement are much crushed and sheared. The Balmacarra Overthrust is cut off under the water of the loch by the greater Moine Overthrust, which comes down to Loch Alsh in the next bay eastward, where it is concealed by a raised beach.

Fig. 42. *Balmacarra Bay*

A, Ard Hill; *B*, Balmacarra; *D*, Donald Murchison's Monument; *l*, Lewisian gneiss of overthrust mass; *t*, Torridonian with beds inverted.

A digression up Loch Duich, which opens on the right, is not without interest. A good place for a landing is Totaig, situated at the sharp bend opposite the narrow mouth of Loch Long. On the shore may be seen in small compass many of the characteristic rocks, Lewisian and pre-Lewisian, of the Glenelg complex: hornblende- and biotite-gneisses, usually with garnet, biotite-amphibolite, good eclogites, and diopside-forsterite-marble. The lower part of the loch traverses what is merely a continuation of the Glenelg country (p. 60), viz. a Lewisian complex including also some bands of the pre-Lewisian rocks, but near its head the Lewisian gives place to the Moine Series, mainly of quartz-schists and granulites. The 'Five Sisters of Kintail' and other mountains which come into view are all carved out of the Moine rocks. The exposures in the lower part of Glen Shiel show some quite coarsely crystalline types, often garnetiferous, invaded at places by pegmatites. Between Ratagan, on the S.W. side of the loch, and the pass leading into Gleann Mòr to Glenelg may be examined the Ratagan plutonic complex, probably one of the Old Red Sandstone intrusions. It consists mainly of varieties of hornblende-syenite, but includes also hornblende- and biotite-granites.

Resuming the passage of Loch Alsh westward to the Kyle, we have Torridonian rocks on both sides. The lower division, inverted, as seen for 2 miles W. of Balmacarra Bay, includes flaggy and shaly beds alternating with grits, some calcareous. About ½ mile before reaching the pier, some flaggy beds have been quarried for slates. Just beyond this come the red felspathic sandstones of the upper Torridonian.

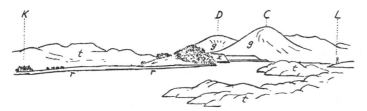

Fig. 43. Seen from the hotel, Kyle of Lochalsh

K, the hotel, Kyleakin; D, Beinn Dearg Bheag; C, Beinn na Caillich; L, Kyle lighthouse; t, Torridonian; g, granite of the Red Hills; r, 30-ft. raised beach; x, 100-ft. raised beach.

Kyle of Lochalsh commands a good view of the nearer part of Skye (fig. 43), and Kyleakin can be reached by ferry (p. 75). From here too one may make an excursion to Balmacarra, Dornie, and Loch Duich or, by road or rail, to Loch Carron. [The railway from Lochalsh to Loch Carron, and an adjoining road branching from the main Balmacarra road 2 miles from Lochalsh, traverse the inverted Torridonian sandstones already noted above, the road striking farther inland till, within 2 miles of Stromeferry, near Fernaig, inverted Lewisian gneiss overlies the Torridonian in a lofty escarpment. At Stromeferry the continuation of the Balmacarra Overthrust brings Moine-like granulites, together with Lewisian gneiss, across highly sheared gneisses with pegmatites and hornblende-schists. In a railway cutting ¾ mile W. of the railway station good exposures show Lewisian pegmatites cut up into lens-shaped masses, a demonstration of the severity of the disturbances in the rocks underlying the overthrust. Near the head of Loch Carron the Moine Overthrust finally introduces the Moine schists, though these are here covered by moraine, a contrast to the bare country to the W. made by the Lewisian gneiss.]

XII. COLL, TIREE, AND THE SMALL ISLES

This chapter will deal with those lesser members of the Inner Hebrides scattered between Mull and Skye. Less frequented than the larger islands, they do not call for so detailed an account, but they have none the less their attractions for the geologist. They fall into two groups: Coll and Tiree, situated to the west of Mull, are composed of Lewisian and pre-Lewisian rocks, and, with the exception of Iona, afford the nearest and most easily accessible representatives of these ancient formations. The second group, lying to the south of Skye, embraces those islands which make up the Inverness-shire parish of Small Isles. Rum is the largest, and grouped about it are Eigg, Muck, Canna, and Sanday. Their geological interest is to be found chiefly in the Tertiary igneous rocks.

(a) COLL

This island, 12 miles in length, is for the most part less than 200 ft. above sea-level. The eastern half is made wholly of Lewisian rocks. Most prevalent are strongly banded grey gneisses, containing horn-blende and biotite, and having a general dioritic composition. Darker bands and lenticles represent enclosed portions of more basic igneous rocks. These are mostly amphibolites (hornblende-plagioclase-rocks), sometimes with augite or garnet in addition, and there is a type composed practically wholly of olivine. On the other hand the gneisses are tra-versed by veins and streaks of coarse felspathic pegmatite. All these varieties may be examined along the road which crosses the island from Arinagour or in the neighbourhood of Grishipoll on the farther coast.

In the western half of Coll the Lewisian gneisses occur as belts with a general N.–S. direction alternating with belts of highly metamorphosed pre-Lewisian sediments. This arrangement is due partly to the manner of intrusion of the igneous gneisses, partly to acute isoclinal folding of the complex. The several belts may be best observed by following the coast or the road south-westward from Grishipoll. At the Church at Clabhach we meet with a narrow band of the old sediments (paragneisses). They have been calcareous sandstones and are composed of dominant

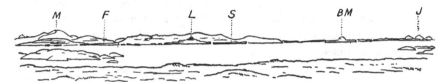

Fig. 44. Looking S.E. from near Breachacha Castle, Coll

M, Ben More of Mull (in front are the Gribun cliffs and the two islands of Cain na Burgh); *F*, Fladda; *L*, Lunga; *S*, Staffa (the Ardmeanach promontory behind); *BM*, Bac Mòr (Dutchman's Cap); *J*, Paps of Jura, seen over the Ross of Mull.

quartz with various lime-minerals—calcite, actinolite, diopside, scapolite, prehnite. There follows a broader belt of igneous gneiss (orthogneiss), rising into Ben Hogh, the highest point of the island. The rock is of pinkish colour, containing biotite and sometimes hornblende, and is of somewhat more acid composition than the commoner grey gneisses.

This side of Coll is in many places covered by blown sands. The next exposures of the pre-Lewisian rocks are to be sought on the shore at the S.W. end of Hogh Bay. They include one of the bands of marble as well as quartzose granulites, some with lime-silicates and others with biotite. The Coll marbles are more or less magnesian, and include such minerals as serpentine (after forsterite), diopside, and phlogopite. Crossing an intervening strip of grey gneiss, we meet the same types, both the highly quartzose and the calcareous on Ben Feall. Here is the broadest belt of the pre-Lewisian rocks of Coll; but this low tract is occupied by a spread of blown sand extending from Feall Bay to Crossapol Bay on the opposite coast. The sand, composed mainly of pounded shells, has often been built up by recrystallization of the carbonate into something like a coherent limestone.

Coll affords a good view of the basalt region of western Mull and the Treshnish Isles (fig. 44).

(b) TIREE

Tiree is of about the same size as Coll and, excepting a couple of hills in the S.W. and W., is equally low-lying. Its geology also is for the most part a repetition of that seen in the sister island, the rocks here shown being similar Lewisian gneisses with intercalated strips of older, much metamorphosed, sediments.

The dominant rock of the island is an acid gneiss, usually containing hornblende as well as biotite, and often veined with pegmatite. Good examples may be seen in Gott Bay near the pier, about Scarinish near the lighthouse, and in many other places.

Of more interest are the Tiree marbles, found especially in the neighbourhood of Ballyphetrish, in the N. of the island. White forsterite- and diopside-marbles, more or less crushed, have been quarried in a field S.E. of the farm; but a more peculiar type, exposed in an old quarry, is a pink marble enclosing crystals of green augite and hornblende. There are nodules of diopside as well as aggregates of large hornblende crystals; and a band composed essentially of diopside is found at the contact of marble and gneiss. Good examples of contacts may be seen near the natural arches on the neighbouring coast, where scapolite has sometimes been produced.

Another occurrence of pre-Lewisian rocks is in the W. of the island, where it makes a narrow belt passing a little W. of Loch Vasapoll and Loch a' Phuill. The principal type here is a garnetiferous gneiss or granulite, representing a band of highly metamorphosed sediments infolded in the midst of the igneous gneiss.

(c) MUCK

This, though the smallest of the group of Small Isles and of rather tame aspect (fig. 36), is by no means devoid of geological interest. It consists almost wholly of Tertiary basalts, with the intrusions which intersect them. However, at the head of Camas Mòr, the principal bay on the south coast, the base of the volcanic series is reached, and Jurassic strata are exposed on the shore (fig. 45). The lowest seen are calcareous sandstones with large concretions, representing the Inferior Oolite. Then comes a considerable thickness of the Great Estuarine Series with leaden-

Fig. 45. Part of the S. coast of Muck, looking W.

This shows the only considerable hill in the island, Beinn Earrair (450 ft.), and the principal bay, Camas Mòr: *e*, Jurassic strata, making reefs between tide-marks; *a*, volcanic agglomerate; *b*, basalt; *d*, some of the many basic dykes.

hued shales alternating with oyster-beds crowded with the little *Ostrea hebridica.* An irregular dolerite sill interrupts the succession, and above is a group of cement-stones with thin shaly bands containing *Cyrena* and a bed rich in *Paludina.*

The lowest part of the volcanic series, resting on the Jurassic beds and exposed near high-water mark on the west coast of the bay, consists of fragmental deposits, volcanic agglomerate with red laminated tuffs below and above. These are succeeded by the regular basalt series. A little E. of Camas Mòr the cliffs are made for some distance by an intrusion of gabbro. It has the form of a large dyke, and can be followed inland nearly across the island, having the same N.W. direction as the ordinary basic dykes.

These last occur in extraordinary profusion, and there is no place better suited for the study of dykes than the isle of Muck. In general they weather out in strong relief. On some parts of the coast, notably in Camas na Cairidh in the north, they present the appearance of a great number of parallel walls running out to sea. They occur on an average at intervals of about 20 yards apart. All are of basic composition, and many of them have selvages of black glass (tachylyte). On the exposed walls the lines of flow are clearly indicated, and it is noticeable that these are often inclined at low angles to the horizontal: in other words, the direction of flow in the dyke-fissure was as much lateral as upward. In the well-bedded shales on the W. shore of Camas Mòr it often happens that the magma has broken away from its vertical fissure to make a small sill along a bedding-plane for a short distance, then resuming the vertical posture. In the forward drive in the horizontal direction the movement was freer in the vertical dyke-fissures than in the connecting sill, and they consequently carried farther. In several instances one may see two neighbouring dykes (really parts of the same dyke), one dying out suddenly upward and the other downward, the connecting sill not being reached by erosion.

Muck possesses a larger proportion of pasture and arable land than most of the isles. This is partly due to a considerable spread of boulder-clay and probably also to another cause. The white sands in the little bays on the north and west coasts are made of comminuted shells, broken down by the surf on the rocky shore; and this sand, carried inland by the wind, has the effect of a natural marling process.

(d) EIGG

Eigg is best known to the casual traveller by the conspicuous ridge of pitchstone named the Sgùrr; but the island is built mainly of the basalt series, with Jurassic strata emerging below along the northern half of the coast-line (fig. 37). There are in addition various minor intrusions.

Among these last may be noticed certain irregular sheets, making reefs which embarrass approach to the island and figure at various points

68

on the coast between the pier and Kildonan. They consist of a basic andesite with a black glassy selvage. This may be seen on the shore of Poll nam Partan, S. of Kildonan, a little bay with sands composed of comminuted shells. The volcanic series includes, besides ordinary basalts, the more felspathic and sodic type mugearite. It is a pale compact rock with a platy fracture, and often contains flattened vesicles occupied by chalcedony. Two flows of mugearite are crossed between Kildonan and the Manse. To see the Jurassic rocks one may visit the seaward slope of the N.E. coast. Most conspicuous are the cliffs made by the Great Estuarine Sandstone, a calcareous sandstone enclosing large concretionary bodies. Below come the Lower Estuarine Shales with *Cyrena* and *Paludina*. They are remarkable for the large number of thin basalt sills intruded at frequent intervals.

The volcanic rocks in the northern part of the island rise to over 1000 ft. Between this and the southern part, culminating in the Sgùrr, is a depression, through which the road leads to the crofter village of Cleadale. On that side the basalt escarpment, much encumbered with landslips, makes a conspicuous amphitheatre. Curiosity will usually take the visitor northward to the little bay named Camas Sgiotaig, famous for its 'musical sands'. The quartz-grains, derived from the neighbouring cliffs, are of uniform grade and partially rounded, and the sand, when dry, emits a shrill sound when trodden. In the larger bay of Laig, to the S., is a stretch of blown sand. It is worth making a digression to Laig Farm, where the mugearite is again seen, and to the coast a little farther W., where, traversed by dykes, is an exposure of fossiliferous Oxford Shales. They contain cordate ammonites, belemnites, etc.

The chief object of interest in Eigg is, however, the Sgùrr, the relic of a thick irregular sheet of pitchstone, showing in many places a very pronounced columnar jointing. Sir Archibald Geikie regarded it as a lava-flow occupying an old valley; but a detailed survey leads to the conclusion that it is intrusive in the basalt series (fig. 46). This indeed is clearly suggested by the steeply inclined surface of contact seen at the eastern end. The structure is really composite, including several sheets of felsite, well seen on the southern escarpment, which visibly cut the pitchstone, but must be closely related to it. The main rock has a black velvety matrix enclosing porphyritic felspar crystals. At the base it becomes more evidently glassy, as may be seen in places on the N. side, but this part is usually decomposed. It encloses little pieces of basalt picked up from the subjacent rocks. At one spot (marked z in fig. 46) it encloses also a trunk of a coniferous tree. There, on the floor of a recess formed by the weathering out of the base of the sheet, the decomposed pitchstone lies transgressively upon a volcanic agglomerate, belonging to the basalt series, which contains small pieces of a different fossil wood. On the

Fig. 46. The Sgùrr of Eigg, seen from the S.W.

b, basalt; p, pitchstone; f, felsite sheets in the pitchstone; yz is the recess beneath the base of the pitchstone sheet and z is the locality of the fossil wood.

S.W. side of the ridge, and following its curve very closely for more than a mile, is a large dyke-like intrusion of porphyritic felsite, which may mark the fissure through which the pitchstone magma rose.

Glacial striae, pointing westward, may be detected in many places on Eigg. There is no regular boulder-clay, but erratics of Moine schists and pegmatite are common. It appears too that at one stage the Sgùrr nourished at least one small local glacier. It flowed down from the most noticeable gap in the S.W. escarpment, and its left terminal moraine, composed of blocks of pitchstone, some of huge size, is seen streaming down to the coast.

The island of Oigh-sgeir (Hyskeir), with its lighthouse, situated 18 miles from Eigg in a W.N.W. direction, is made by a pitchstone identical with that of the Sgùrr and probably once continuous with it.

(e) RUM

This, the largest of the Small Isles, possesses some geological features of special interest. The northern half of the island, rising gradually to 1000 ft., is a rather monotonous tract of reddish felspathic sandstones with pebbly beds, the familiar Torridon Sandstone. Torridonian strata extend also down the eastern side of the island to the southerly point; but here a group of shales appears beneath the sandstones, making all the south-eastern coast. All the rest is of Tertiary igneous rocks, viz. some rather scanty relics of the volcanic series and a plutonic complex which builds the southern mountainous half of Rum, rising to no great altitude but presenting a bold mountain outline.

The Torridonian strata of this southern area have been affected by an overthrust belonging to the great Caledonian system of crustal displacements; but, except on the N.E. and S.E. sides, the evidence has been obscured by the subsequent plutonic intrusions. The chief effect was to carry large masses of the (lower) shales over the (upper) sandstones. With this there is much violent folding set up and a production of a considerable thickness of crush-breccia bordering the surface of movement.

The volcanic rocks of Rum include basalt lavas, with interbedded fluviatile conglomerates, and rocks approximating to the mugearite type, having oligoclase as their dominant felspar. The plutonic complex is remarkable among all British occurrences for the prominence and fresh preservation of ultrabasic types. Besides ordinary peridotites there is found the pale felspathic type allivalite, essentially an olivine-anorthite-rock. On some mountain slopes this is seen to make thick sheets alternating with sheets of peridotite. Next in order came eucrites, intruded beneath and into the ultrabasic group, and finally granites. A remarkable feature is the production of well-banded gneisses by flowing movement set up in what was a half-digested mixed rock made from eucrite and granite. There are also intrusions of porphyritic quartz-felsite in irregular sheets. Basic dykes represent the latest igneous episode

71

Although Rum is familiar in a distant view to most travellers among the western isles, it affords no facilities for the ordinary tourist, and we shall be content to notice its more striking features as observed from a distance. Some of these are well displayed as the island is approached from the S.E. (fig. 47). The effect of the overthrust is most clearly seen

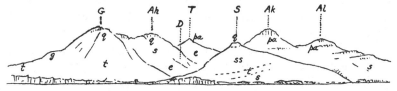

Fig. 47. Rum, seen from the S.E.

G, Sgùrr nan Gillean; Ah, Ashval; D, Glen Dibidil; T, Trallval; S, Beinn nan Stac; Ak, Askival; Al, Allival; s, Torridonian shales; ss, shales above the overthrust; t, Torridon Sandstone; pa, stratiform alternations of peridotite and allivalite; e, eucrite; g, granite; q, porphyritic quartz-felsite.

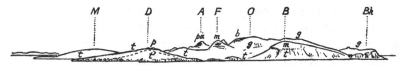

Fig. 48. Rum, seen from the N.W., from Canna

M, Mullach Mòr; D, Monadh Dubh (the summit of Sgaorishal, made by a small intrusion of peridotite, appears behind); A, Allival; F, Fionn-Chrò; O, Orval; B, Bloodstone Hill (Creag nan Stardean); Bh, A' Bhrideanach; t, Torridon Sandstone; t', brecciated sandstone; m, mugearite-basalt; b, basalt; p, peridotite and pa alternations of peridotite and allivalite; g, granite.

on Beinn nan Stac, E. of Glen Dibidil, the seaward slope of which is made up mainly of highly contorted shales above the surface of displacement, which has here an inclination not very different from the slope of the hill. In Askival and Allival the strong escarpments are made by the allivalite, the gentler slopes by the peridotite. The eucrite makes a pronounced platform below. The upper part of Sgùrr nan Gillean and Ashval is a thick sheet of quartz-felsite, and the same rock, intruded in a crush-breccia, caps Beinn nan Stac.

Other features of interest are brought out in a distant view of Rum from the opposite (N.W.) side (fig. 48). The seaward face of Monadh Dubh is made by a relatively thin cake, mainly of crush-conglomerate of sandstone, carried on a surface of overthrust. It is doubtless in inverted position beneath some more powerful overthrust, for the part immediately overlying the surface of movement consists of broken and crushed Cambrian limestone, a formation not otherwise represented on Rum. In the Bloodstone Hill we see a dip-slope of Torridon Sandstone and resting on it nearly horizontal flows of mugearite-basalt. The same type of lava makes the summit of Fionn-Chrò, where it overlies basalt flows. An outline of metamorphosed basalt clings to the shoulder of the granite mountain Orval. The granite which makes A' Bhrideanach, and extends southward, has a pronounced jointing in vertical columns.

(f) CANNA AND SANDAY

Canna and its smaller neighbour Sanday, near enough to be joined by a footbridge at low tide, may be considered together. As might be inferred from the terraced appearance of the hills, very noticeable on the larger island, the Tertiary basalts make the chief element in the geological constitution (fig. 49). No older rocks are exposed, but there are indications that the Torridonian is present not far below sea-level, the Mesozoic strata (if ever deposited in this area) having been removed by pre-Tertiary erosion. What is probably the basal portion of the volcanic succession, as exhibited in the eastern part of Canna and Sanday, has a special interest from the development here of a thick group of fragmental deposits.

A good section is to be seen in Compass Hill, on the E. coast of Canna. The lowest member exposed is a coarse agglomerate 100 ft. in thickness, containing subangular blocks of basalt up to 6 ft. in diameter. There are also many pieces of Torridon Sandstone, but no Jurassic rocks are to be recognized. The finer matrix is of basaltic composition. Resting on this, and probably derived from it is an ashy conglomerate or conglomeratic tuff, well bedded and carrying rounded pebbles. It has doubtless been laid down by a river flowing over the agglomerate. Two more intercalations of bedded tuffs with pebbles occur in the upper half of the cliff, and then follows the pile of basalt flows which makes the bulk of the island. About the summit the rocks are highly magnetized in a manner to cause great disturbance of the compass: hence the name of the hill. The phenomenon is, however, in no wise peculiar. Prominent points on the basalt moorland, not only in Canna but in Skye and elsewhere, very generally show strong permanent magnetization.

South of Compass Hill, at the two stacks named Coroghan Mòr and Alman, one may see a good illustration of the resistance offered by coarse fragmental deposits to sill-intrusions. Along the shore of the Harbour the

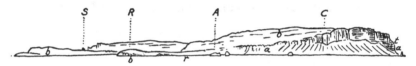

Fig. 49. Canna and Sanday, seen from the E.

S, Sanday Catholic Church; *R*, Rudh Carr-innis, guarding Canna Harbour; *A*, Alman, with Coroghan Mòr to the right; *C*, Compass Hill; *a*, volcanic agglomerate and fluviatile conglomerate; *t*, bedded basaltic tuff; *b*, basalt; *r*, raised beach.

agglomerate continues, in places evidently rearranged by water transport, but the upper part of the Compass Hill section is represented only by inconstant beds of tuff, sometimes with pebbles. A good exposure is seen close to the first cottage W. of Canna House, where one band contains plant remains. Along the coast westward are beautiful examples of columnar jointing in the basalts. The fragmental group no longer appears above sea-level; but it is probable that the basal agglomerate persists for some distance westward, and to this one may probably attribute the paucity of dykes in these islands.

There is no continuous spread of boulder-clay on Canna and Sanday, but erratic boulders are common. They are of Torridon Sandstone with eucrite and granite (probably from Rum) and various Moine rocks from the mainland. High raised beaches occur at the E. end of Canna and at Tarbert, on the narrow waist of the island.

The small islands in Canna Harbour show the agglomerate, and in one case it is directly overlain by a mugearitic rock like those of Rum (p. 73). It makes the lower part of a flow, of which the upper half is an ordinary basalt.

On Sanday the fragmental volcanic deposits are much reduced. Round the bay where the crofts are situated only bedded tuffs are found, and the coarse agglomerate is last seen between tide-marks at the E. end of the island.

XIII. ROUND THE ISLE OF SKYE

Skye, the largest of the Inner Hebrides, presents a considerable variety of geological constitution. In the S.E. the Sleat district, as we have seen (pp. 58–60), is composed of some of the oldest rocks. The adjacent district of Strath, S.W. of Broadford, is made largely by Torridonian and Cambrian, but there is also a larger spread of Mesozoic strata than is found elsewhere in the region, besides numerous Tertiary intrusions. The central part of the island, built essentially of Tertiary plutonic rocks, possesses most general interest. Here are the Cuillin and Blaven ranges, the finest of all British mountains, composed mainly of gabbro, and contrasting with these the smoother outline of the granite 'Red Hills'. To the N.W. is the largest continuous extent of basalt-plateau country, rising to over 2000 ft. in the bold escarpment on the E. coast, with Jurassic strata emerging below. Only some of the salient points of the geology of Skye can be noticed here. According to our plan, we shall circumnavigate the island, halting at the chief places of resort.

(a) KYLEAKIN AND BROADFORD

At Kyleakin, the principal interest is in the raised beaches (see fig. 43). The higher beach, like a great level embankment, makes a striking feature, and there is a good section of it in the little ravine made by the

75

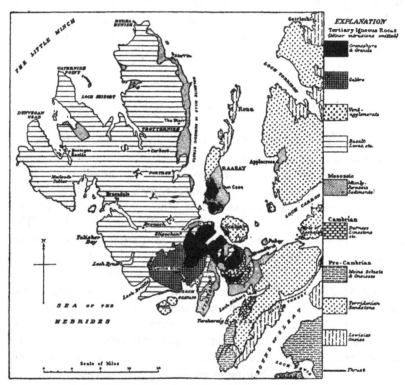

Map III. Geological Sketch-Map of Skye

[Reprinted from Fig. 4, British Regional Geology: Scotland: The
Tertiary Volcanic Districts (*Mem. Geol. Surv.*) 1935. By permission
of the Controller of H.M. Stationery Office.]

Anavaig River. A curious detail is the Ob, a salt-water creek running inland behind the beach and filled at high tide. Resting on Torridon Sandstone, the raised beach can be followed along the coast-road halfway to Broadford. Westward there is a good view of some of the Skye mountains (fig. 50).

Broadford is more usually approached by water. As we leave the Kyle, the view opens out to W. and N., showing the islands of Raasay, Crowlin, etc. (see fig. 68). Nearing Broadford, we pass the low flat island of Pabay, a locality for Lower Lias fossils. This is a good point for a general view

Fig. 50. Looking W. from Kyleakin

DB, Beinn Dearg Bheag and *C*, Beinn na Caillich, the Red Hills of Broadford; *B*, Blaven and *SG*, Sgùrr nan Gillean, gabbro mountains; *BC*, Beinn na Crò, with summit of granite and N. ridge of metamorphosed basalt; *GM*, Glas Bheinn Mhòr, granite; *CS*, Creag Strollamus, the upper part of granophyre; *D*, Beinn Dearg, the Red Hills of Sligachan; *G*, summit of Glamaig, metamorphosed basalt; *S*, Scalpay, of Torridon Sandstone; *P*, Pabay, Lias. The foreground is on the high raised beach.

of the local geology (fig. 51). Ben Suardal is on the axis of a sharp anticline, which however is dying out in this direction. Its core is the Cambrian limestone which makes the summit ridge, the Torridon Sandstone dipping off from it on both sides. It is to be observed that, in consequence of an overthrust (older of course than the anticline) the Torridonian of this district *overlies* the Cambrian. The Mesozoic beds—Lias with a Triassic conglomerate at the base—resting with strong unconformity upon the Torridonian are also affected by the anticline, but sweep round the end of it, making all the coastal tract. Westward a fault brings on the Cambrian limestone again. Here it is invaded by Tertiary gabbro and also by the granite of the Red Hills. On the lower slope of Beinn Dearg Bheag, between Ben Suardal and Torran, a stretch of rough ground marks the place of the large volcanic vent of Kilchrist, and in the distance is seen part of the outlier of basalt in the Strathaird district.

The Broadford crofts, like most others, are situated on a raised beach. The Lias is exposed along the coast, intersected by numerous Tertiary N.W.–S.E. dykes, often standing up in relief. The beds exposed in front of the village and eastward belong to the lower division of the

Fig. 51. *Approaching Broadford, from near Pabay*

S, Ben Suardal; *A*, An Càrnach; *D*, Beinn Dearg Bheag; *C*, Beinn na Caillich; *CD*, Creagan Dubha; *t*, Torridon Sandstone; *c*, Cambrian limestone; *l*, Lias; *a*, volcanic agglomerate of the Kilchrist vent; *b*, basalt (that at Creagan Dubha metamorphosed by the granite, but faulted against it); *gb*, gabbro, intrusive in the Cambrian limestone; *g*, granite of the Red Hills; *r*, raised beach.

Lower Lias (Broadford Beds), and include dark blue limestones, sometimes pebbly, and calcareous sandstones. Fossils are abundant, and some beds W. of the river are made up largely of *Gryphaea arcuata*. On the W. side of the bay are some interesting intrusions. A little beyond Corry Lodge two dolerite dykes on the shore contain so much debris of granite, partly digested, that their composition has been much modified. There is a good example too of a multiple dyke, comprising a number of successive injections into the same fissure. The headland Rudh' an Eireannaich at the N. end of the bay is made by an interesting composite sill. It has upper and lower members of porphyritic basalt with a central intrusion of pale bostonite, and there is a gradual transition from the basic to the felspathic rock. A little beyond this are one or two sills of a beautiful spherulitic granophyre. These are intruded in the Pabay Shales, the upper division of the Lower Lias.

A short excursion along the Heast road should be made to examine a group of composite sills intruded in the Lias. They have borders of basalt with an interior of granophyre, which has attacked the basic rock at the surfaces of junction. Examples are seen on the road itself, and a larger one is clearly exposed on the hill Cnoc Carnach.

A very interesting day may be spent along the Torran road. Starting up a tributary valley, it runs for more than a mile on Lias, with Torridon Sandstone seen not far to the left. Where the road makes a steep drop, it crosses the Trias conglomerate, with pebbles of Cambrian quartzite and limestone. Descending into the main valley (Strath Suardal), it is possible to verify, by the roadside and in the abandoned light railway above, the super-position of Torridon Sandstone on Cambrian limestone brought about by the overthrust (p. 77). Note too, in the railway, how Tertiary basalt dykes, cutting the limestone, are stopped by the more intractable sandstone above. The limestone here is the upper of the two zones recognized in this district, a pale rock with numerous yellow worm-casts and bands of small black cherts. Fossils are not easily found. On the west side of the road (which follows a line of fault) and on the floor of the strath the lower zone may be seen, a dark saccharoidal dolomite enclosing many siliceous sponges. The road, however, continues on the upper limestone, and follows the shore of Loch Kilchrist. A ridge abutting upon the loch is made by a large dolerite dyke. It belongs to a pre-granite set, and on following the ridge up the slope it is seen that the dyke, with a bordering strip of limestone, becomes broken and enveloped by the granite mass which, farther on, makes Beinn an Dubhaich. The dolerite is metamorphosed with the production of hornblende. Along the hillside southward numerous patches of limestone have been enclosed in the granite, which has cut through vertically without disturbing the bedding. The limestone is metamorphosed to a forsterite-marble, and the forsterite afterwards often changed to yellow serpentine. It belongs to the sponge-bearing zone, and there are curious concentric shell structures round the vanished sponges. Note that the

79

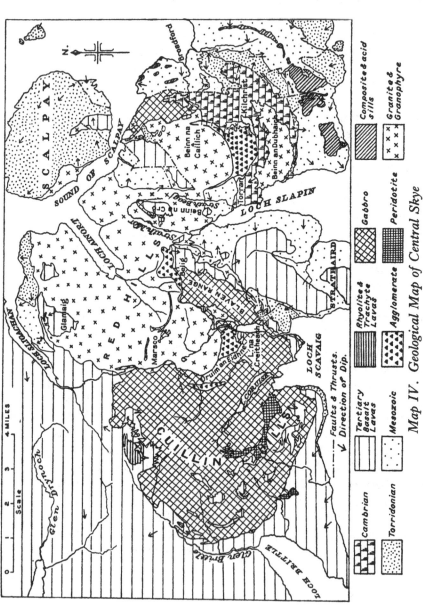

Map IV. Geological Map of Central Skye

Redrawn from coloured map by A. Harker in 'The Tertiary Igneous Rocks of Skye' (*Mem. Geol. Surv.*) 1904. [Reprinted, with permission, from Fig. 8, 'Tertiary Ring Structures in Britain', *Trans. Geol. Soc. Glasgow*, vol. XIX, pt. I, 1931-32.]

granite of Beinn an Dubhaich is intruded along the prolongation of the Ben Suardal anticline, which, like the Mull axes (p. 49), has a curved course. The granite differs from that of the Broadford Red Hills in carrying hornblende as well as biotite. Returning to Loch Kilchrist, the geologist should examine the large Kilchrist volcanic vent, which occupies the ground N.W. and W. of the loch, being cut off to the N. by the granite of the Red Hills. It is filled by a basaltic agglomerate with blocks of basalt and sandstone. Intruded along its margin are several occurrences of a granophyre full of half-digested inclusions of gabbro, well seen on the track which skirts the farther shore of the loch. By

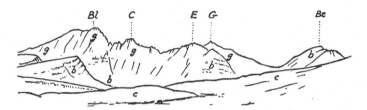

Fig. 52. The Blaven range, seen from near Kilbride

Bl, Blaven; *C*, Clach Glas; *E*, Sgùrr nan Each; *G*, Garbh-bheinn; *Be*, Belig;
c, Cambrian dolomitic limestone; *b*, Tertiary basalts; *g*, gabbro.

following the high-road as far as Kilbride one may gain a good view of the Blaven range to the W., showing the large gabbro laccolite overlying the basalt lavas (fig. 52).

Another interesting walk may be taken along the Sligachan road. About a mile out this meets the fault throwing the Lias against the Cambrian limestone, and follows it for some distance. The limestone territory on the left, however, is largely occupied by a gabbro intrusion. The relations are like those already noticed for granite and limestone (p. 79), numerous patches of marble being enclosed in the gabbro. On the shore in front of Creag Strollamus, between two faults, is a strip of granite so thoroughly brecciated as to resemble an agglomerate. Farther on, before reaching Strollamus Lodge, glaciated crags of basalt are seen at the roadside. Along all this part of the road Scalpay is in full view. It is essentially a Torridon Sandstone island, but the S.E. corner, where Scalpay House stands, is made by Middle Lias beds faulted down. Farther W. is a faulted inlier which includes indurated shales of Oxfordian age and a narrow strip of Cretaceous beds, covered by the Tertiary basalts (fig. 53). It may be remarked that the dividing line between the native and foreign ice-sheets ran along the Sound of Scalpay. On that island, which was overrun by the Scottish ice, boulders from the mainland occur at all altitudes, but on the Skye side they are found only in the beaches.

Fig. 53. *Scalpay, seen from Strollamus Lodge*

M, Mullach nan Càrn; *t*, Torridon Sandstone; *o*, Oxfordian shales and sandstone; *c*, Cretaceous limestones and sandstones; *b*, Tertiary basalts; *f*, fault.

(b) BROADFORD TO PORTREE

The Kyle of Scalpay not being navigable, the steamer bound for Portree must pass outside the island, where the reddish Torridon Sandstone shows a regular dip to N.W. In front is Raasay, presenting in its long eastern escarpment a complete section of the Lias, surmounted by the Inferior Oolite and Great Estuarine Series, and capped by an outlier of the Tertiary basalts. Passing between the two islands, note on the north-west corner of Scalpay a small outlying patch of Triassic conglomerate. The Skye coast beyond is made by a strip of Torridon Sandstone but behind is the large granite tract of the main Red Hills (fig. 54). The upper part of Glamaig is of basalt lavas, nearly surrounded by granite. We look into the mouth of Loch Sligachan, but Sligachan itself is more naturally reached from Portree. North of this loch begins the large area of basalt lavas which occupies all the N. and N.W. of Skye.

The signs of industry at the S. end of Raasay are accounted for by the occurrence of a bedded ironstone in the Upper Lias of this district. It is an oolitic ore of the chamosite type. The slopes above are composed of a riebeckite-bearing granophyre, intruded in the Jurassic rocks, and the low headland on the Skye side of the Narrows is also of granophyre. On the W. side of Raasay this rock comes down to the coast, until farther N., opposite Portree Bay, a fault brings on the Torridon Sandstone. The basalt lavas on this part of the Skye coast belong to the base of the thick succession, and members of the Oolite Series emerge below, rising to 1000 ft. on the side of Ben Tianavaig.

Fig. 54. Seen from the N.W. corner of Scalpay

M, Marsco; D, Beinn Dearg; G, Glamaig; t, Torridon Sandstone; c, Trias conglomerate; b, Tertiary basalts; g, granite.

(c) PORTREE

Portree is a suitable place for examining the base of the volcanic succession and its relations to the subjacent rocks (fig. 55). By crossing the harbour one may inspect one of the accumulations of agglomerate and tuff which in not a few localities mark the explosive character of the earliest volcanic outbreaks. The view northward shows the long range of the basalt escarpment, culminating in the Storr, with the conspicuous 'Old Man' standing out in front. The seaward slope is made by the Jurassic strata, but there is some complication caused by landslips. Southward the road runs through Glen Varragil, with typical basalt-plateau scenery to Sligachan, a place which calls for more particular notice. The broad valley is largely drift-covered, and a considerable tract is studded with little hillocks resembling tumuli. This is a characteristic example of the 'hummocky drift' or 'kettle moraine' due to the melting of a stagnant sheet of ice.

Portree is also a convenient starting-point for a visit to Raasay (p. 83). North of the landing-place, between the Chapel and Oskaig Point, is a sill of a coarse crinanite. The main road northward runs for about three miles on riebeckite-bearing granophyre. Soon after, it crosses the fault which throws down the Jurassic rocks (Inferior Oolite) against the Torridon Sandstone. By continuing as far as Brochel Castle one may examine the shaly base of the Torridonian, resting on Lewisian gneiss. The ruined castle stands on a small volcanic neck of Tertiary age.

Fig. 55. The N. side of Portree Harbour, seen from the Hotel

V, Vriskaig Point; *T*, Ben Tianavaig; *S*, Scor farm (here is a landslip; the ravine beyond is made by a dyke); *l*, Middle and Upper Lias; *o*, Inferior Oolite sandstone; *a*, volcanic agglomerate; *b*, basalt lavas.

(d) THE CUILLINS AND RED HILLS FROM SLIGACHAN

Sligachan, the starting-point for climbers in the northern Cuillins, is also a centre of much geological interest. We shall note here a few of the more important features. The mountains are seen across an intervening tract of basalt lavas, mostly covered by superficial deposits. Most striking is the contrast between the broken ridges of the Cuillins, essentially gabbro mountains, and the smooth outlines of the granite Red

Hills (fig. 56). The difference is largely due to the multitude of dykes cutting the gabbro, which weather into notches and ravines, while the few dykes in the granite hills tend to stand out in relief.

Glamaig, the nearest mountain towards the E., is situated on the edge of the extensive tract of granite. Its summit ridge is of basalt, more or less metamorphosed, with an intrusion of peridotite, and the northern slopes are mainly of Torridon Sandstone and Lias; but the side towards Sligachan shows little besides granite. The shoulder of the Red Hills named Sròn a' Bhealain, S.E. of Sligachan, is made by a sheet of a peculiar rock

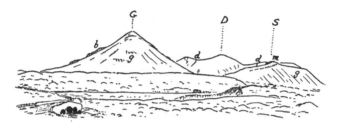

Fig. 56. Glamaig and the Red Hills, seen from Sligachan

G, Glamaig; D, Beinn Dearg (the gorge of Allt Daraich seen in front); S, Sròn a' Bhealain; b, basalt lavas; m, marscoite sheet; g, granite; d, basalt dykes. The nearer ground is of basalt covered by drift and peat.

known as marscoite, a hybrid of gabbro and granite, and the sheet has been enveloped and attacked by the granite magma. To see this, go up the glen of Allt Daraich, the lower part of which is a conspicuous gully made by a multiple dyke with pronounced hade (fig. 56). About a mile up the basalt gives place to granite, and half a mile farther abundant relics of the marscoite sheet are seen in the burn. One type is a spotted rock made by the granite or granophyre enclosing little round patches of marscoite, and further reaction has given rise to a hybrid of the two rocks.

The gabbro tract of Skye includes the Cuillins and the Blaven range, one continuous tract but encroached on northward by the granite, intruded later, and in a general way beneath the gabbro. The great

85

gabbro mass, made up of numerous separate intrusions, has the habit of a lens or laccolite. The basalts are seen passing beneath it, and outliers of the volcanic rocks on the highest summits probably represent the roof of the laccolite. There are also lenticles of basalt, much metamorphosed, entangled in the gabbro. In all the upper parts of the mountains the gabbro is intersected by an immense number of inclined sheets of dolerite. The inclination is inwards, and in these northern mountains the sheets have therefore a southerly dip. In addition there are very numerous basic dykes, mostly with the regular N.W.–S.E. direction.

Fig. 57. Outlines of Sgùrr nan Gillean

The low ridge of Nead na h'Iolaire (*I*) marks the boundary of the gabbro tract. *G*, Sgùrr nan Gillean (the 'Pinnacle Ridge' running out directly towards Sligachan is not clearly seen); *B*, Basteir Rock, with Coir' a' Bhasteir in front; *SB*, Sgùrr a' Bhasteir.

In general they are cut by the inclined sheets. The easiest way of observing all these features is by a visit to Coir' a' Bhasteir, approached from Sligachan by Allt Dearg Beag (fig. 57).

An excursion up Glen Sligachan affords numerous observations of geological interest. In the lower reach there is a covering of hummocky drift and peat, with fluviatile gravels bordering the river. For about $1\frac{1}{2}$ miles the track is on basalt lavas, somewhat indurated and with veinlets of epidote. After this we have gabbro on the right and granite on the left, the junction following the trend of the valley. The track is for the most part over granite, but after passing the boundary fence it runs for some distance along a strip of gabbro enclosed in the granite near its margin. Farther on it encounters a boggy tract made by a delta of red sand washed down from a deep gully on the N.W. flank of Marsco. This is due to a strip of gabbro, of dyke-like habit, caught in the granite and attacked by it, the resulting hybrid product being always an easily destructible rock. On the N. side there is a narrow border of a hybrid rock 'marscoite'. A prominent shoulder of Marsco shows a pronounced columnar jointing. In the river near here the almost vertical junction of gabbro and granite can be verified. As we approach the watershed, near the two small lochs, the Blaven range comes well into view, showing the granite passing under the gabbro (fig. 58). Taking the tourist track to

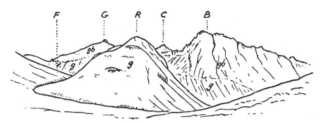

Fig. 58. Ruadh Stac and Blaven, seen from the watershed at Loch Dubh
F, Am Fraoch-choire; G, Garbh-bheinn; R, Ruadh Stac; C, Clach Glas; B, Blaven; gb, gabbro; g, granite.

Loch Coruisk, which branches off from the Camasunary path, we are first on granite, but the upper half of the ascent is over a large wedge-like area of volcanic agglomerate, caught between gabbro and granite. The ridge itself, Druim an Eidhne (Drum-hain) is made by the border of the main gabbro tract. From the pass there is a fine panorama of the western Cuillins, and the track leads down to Loch Scavaig and the outlet of Loch Coruisk (p. 94). The geologist, however, will find it interesting to return along the ridge, north-westward. In this direction the agglomerate tails out, bringing gabbro and granite together. The relative age of the two is evident; for the granite assumes here a compact texture like a rhyolite, and sends out offshoots in the form of dykes, penetrating the gabbro and having a beautiful spherulitic structure. The gabbro in this neighbourhood shows some remarkable examples of a strongly-banded structure. Darker and paler bands, more augitic and more felspathic, alternate, and there are black seams very rich in titaniferous iron ore. The granite also presents points of interest. In the Red Hills generally it shows some variety, due doubtless to distinct intrusions. Here, as we approach the hill Meall Dearg, which guards the mouth of Harta Corrie, we find a riebeckite-granophyre, while the descent of Meall Dearg itself is over a typical augite-granophyre.

During the maximum glaciation when the Skye mountains were buried under a local ice-cap, a powerful body of ice flowed down Glen Sligachan, gathering boulders on its way. As might be expected, these are mainly of granite on the right (E.) side and gabbro on the left (W.), the median line, where gabbro and granite are equally abundant, following nearly the course of the river. This line can still be traced when, lower down the glen, basalt boulders come to preponderate more and more over gabbro and granite. By observing this it can be verified that at the mouth of the glen, on encountering the great Scottish ice-sheet which pressed on the E. coast of Skye, the Glen Sligachan stream was forced westward, over hill and valley round the Cuillins.

The Tertiary volcanic rocks in this neighbourhood are not exclusively of basaltic nature. On the northern flanks of the Cuillins there have been local eruptions of trachytic and rhyolitic composition. These may be seen by following the Glen Brittle footpath. The track itself, along Allt Dearg Mòr, S.W. of Sligachan, is on basalt, but the more acid types, much cut up by tongues of gabbro, are exposed on the slopes to the south, between Meall Odhar and the N. ridge of Bruach na Frithe. The two principal streams here are Allt an Fhionn-choire and Allt Mòr an Fhinn Choire. In the former are exposed greenish fine-textured trachytes with intercalated flows of a porphyritic andesite. Between the two streams one may see rhyolitic tuffs and agglomerate, and in the upper part of the second burn named the rhyolites are conspicuous, usually a good deal altered, with strings of secondary quartz. This is roughly the order of succession of this local development, which comes in the midst of the basalt lavas, into which it dovetails on both sides. Note, along the Glen Brittle track, many fragments of slaty rhyolite, white or pale mauve. These have been brought down at a time when the main glaciation had been succeeded by valley-glaciers following the natural drainage-lines.

(e) ROUND NORTHERN SKYE

We resume now our circumnavigation of the Isle of Skye. The cliffs N. of Portree present the same characters as before. On the other side of the Sound the low-lying northern part of Raasay and the island of South Rona beyond show an ice-moulded tract of Lewisian gneiss. A feature of this part of the Skye coast is the occurrence of numerous dolerite sills intercalated in the Jurassic strata, sometimes slightly shifting their

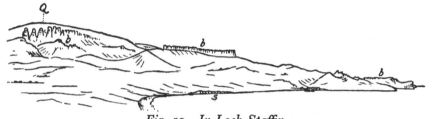

Fig. 59. In Loch Staffin

Q, the Quiraing; *s*, Oxfordian shales; *b*, basalt lavas.

horizon. There are some striking examples of columnar jointing. Successively higher members of the Jurassic come into the cliffs as we proceed, until in Loch Staffin the Oxfordian shales are seen at sea-level (fig. 59). Above here the long range of basalt hills, which from the Storr has receded somewhat inland, terminates in the Quiraing,[1] with its remarkable assemblage of pinnacles. This is a good place for examining the amygdaloidal basalt lavas, which contain numerous species of zeolites.

In the neighbourhood of Rudha Hunish, the most northerly point of Skye, and in the rocky islets seen beyond, outliers of dolerite sills are

[1] [The Quiraing belongs to a great series of landslips which extend all along the eastern escarpment of the basalt plateau and give rise to the irregular and rough scenery of this part of Skye.]

responsible for all the salient features. The N.W. coast of the island, with the deep indentations of Loch Snizort and Loch Dunvegan, does not call for detailed notice. All is typical basalt-plateau scenery, though Jurassic strata emerge at a few places. These are seen again after passing the Eist lighthouse and the conspicuous cliff of Waterstein (fig. 60). Farther on we pass Idrigill Point with the basaltic stacks known as Macleod's Maidens, and then the wide mouth of Loch Bracadale (fig. 61). Looking back, we have a view of the two flat-topped hills named Macleod's Tables, carved out of horizontal flows of basalt.[1]

Fig. 60. Coast of Skye near Eist

L, Eist lighthouse; e, Great Estuarine Series; b, basalt lavas.

Fig. 61. Seen from Loch Bracadale looking N.

I, Idrigill Point; HB, Healaval Beg; HM, Healaval More; W, Wiay island; b, basalt lavas.

On Loch Harport, the eastern branch of Loch Bracadale, is Carbost, which is a port of call for steamers or can be reached from Portree and Sligachan. It is a convenient centre for examining this district, and is within easy reach of the north-western Cuillins.

Soon after passing Loch Bracadale we look into Talisker Bay. The hill Preshal Mòr, rising behind Talisker House, shows a fine example of columnar jointing (fig. 62). Talisker can be reached by road from Carbost. The cultivated ground is made, as usual, by a raised beach. From the

---

[1] [Among the lavas of Macleod's Tables a few occurrences of Tertiary sediments have been recently noted. One of them includes a good plant-bed containing well-preserved leaves of deciduous trees.]

fallen blocks of amygdaloidal basalt below the surrounding cliffs numerous species of zeolites may be collected. Beyond Talisker Bay the precipitous cliffs, rising over 900 ft., afford a good section of the basalts. A noticeable feature is the occurrence of bands of red clay or bole, due to the subaerial weathering of a flow before it was covered by its successor. The coast scenery from here to Loch Eynort and beyond offers no new points of

Fig. 62. *Talisker Bay*

R, Rudha Cruinn; T, Talisker House; P, Preshal Mòr; S, Talisker Point with stack.

interest, but, as we approach Loch Brittle, the main range of the Cuillins fixes attention, and calls for a digression at this point.

(f) THE CUILLINS FROM GLEN BRITTLE

Glen Brittle, the starting-point for climbing on the western mountains, is equally a point of vantage for the geologist, but we can afford space only for a summary notice. There is a comprehensive view of the western

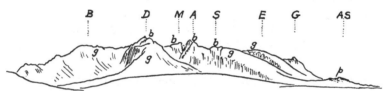

Fig. 63. *The Western Cuillins, seen from Glen Brittle.*

B, Sgùrr na Banachdich; D, Sgùrr Dearg; M, Sgùrr Mhic Coinnich; A, Sgùrr Alasdair; S, Sgùrr Sgùmain; E, Sgùrr nan Eag; G, Gars-bheinn; AS, An Sgùman; b, basalt lavas with some agglomerate; g, gabbro; p, a late intrusion of peridotite.

mountains (fig. 63). The cultivated ground on the floor of the valley is made by raised beaches, rising to 100 ft. or more. Above this the basalt series is exposed for some 200–300 ft. up the slope before the gabbro is reached. Considerable strips of the volcanic rocks are enclosed also in the gabbro, and the summit-ridge of the highest peaks is composed of the same rocks in a metamorphosed state, perhaps indicating the roof of the large gabbro laccolite. On the higher slopes the inclined sheets of dolerite (p. 86), with inward dip, are seen outcropping in great numbers. Basic dykes are numerous, some older, some younger than the sheets. There are also some ultrabasic dykes, recognizable by their reddish weathering crust, conspicuous examples being seen on Sgùrr Dearg.

The volcanic series includes some interbedded fragmental deposits. There are good exposures in two small burns 200–300 yards N.E. of Glen Brittle House. Here is seen conglomerate with intercalated beds of fine tuff. The conglomerate is cut off and metamorphosed by the gabbro, as seen near the footbridge over the Banachdich burn.

The Cuillin range presents a striking picture of the effects of ice-erosion, the work in the main of the ice-cap which covered all the mountain tract, the summit-ridge acting as ice-shed. The chief scenic features to note are: the widening and straightening of the valleys by the grinding away of all lateral projections, giving the characteristic U-shaped cross-section; the excessive erosion at the head of each valley, which has been developed into a 'cirque' or true corrie; the extreme acuteness of the main ridge and the upper parts of branch ridges, thus cut back, and the cuspate form of the peaks; the irregular gradient of the valleys in longitudinal section, giving in places a steep drop, once an ice-fall, and sometimes a small lake-basin.

The material eroded from the mountains is spread as drift deposits over the lower ground. Besides the regular ground-moraine there are areas of hummocky drift (p. 84). The great screes which are a feature of the corries are also of late-Glacial origin, due to frost-weathering of the ridges, which continued after the glaciers (the latest phase of the ice-age) had vacated the valleys. The contrast between the shattered ridges and the smooth slopes below is very noticeable. In Coir' a' Ghrunnda, the highest of the corries, the ice lingered longer than elsewhere, and at its mouth is a fine horseshoe moraine, answering to the screes in the other corries.

(g) ROUND SOUTHERN SKYE

We resume now our circumnavigation of Skye. The peninsula which terminates in Rudh' an Dunain is built of the basalt succession. Rounding it, we enter the Sound of Soay and look into Coir' a' Ghrunnda and Coire nan Laogh. The low island of Soay on the right consists of felspathic sandstones of the Torridonian, intersected by numerous minor intrusions of Tertiary age. On the Skye side the base of the basalt series presently rises above sea-level. Between this and the underlying Torridonian occur scanty representatives of Trias, Lias, and Cretaceous. These are

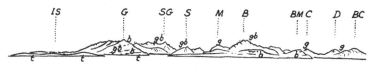

Fig. 64. Looking into Loch Scavaig

IS, Island of Soay; *G*, Gars-bheinn; *SG*, Sgùrr nan Gillean; *S*, Sgùrr na Stri; *M*, Marsco; *B*, Blaven; *BM*, Ben Meabost; *C*, Beinn na Crò; *D*, Beinn Dearg; *BC*, Beinn na Caillich; *t*, Torridon Sandstone; *b*, basalt lavas; *gb*, gabbro; *g*, granite.

little in evidence, the only conspicuous feature being the little promontory, composed of Triassic conglomerate, used as a landing-place by the Soay fishermen.

We come next to Loch Scavaig. This, with its backing of mountains, is best seen from the south, entering or leaving (fig. 64). Entering from the west, round the base of Gars-bheinn, we see the basalt lavas sink again below sea-level, giving place to highly glaciated gabbro. The earliest plutonic intrusion of the Cuillins was of peridotite, making an irregular laccolite, afterwards enveloped and attacked by the more voluminous gabbro. It makes part of the Sgùrr Dubh ridge and the valley named An Garbh-choire (fig. 65). On the shores of Loch Scavaig and Loch Coruisk the gabbro encloses many fragments of the peridotite.

A walk of a quarter of a mile from the landing-place at the head of Loch Scavaig leads to the foot of Loch Coruisk. The lake lies in a rock-basin 125 ft. deep, and so reaching 100 ft. below sea-level. There is a fine view of the mountains, intersected by many inclined sheets, which here dip downward, offering less invitation to the climber than on the

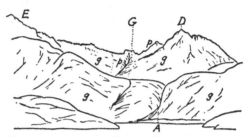

Fig. 65. From the boat-landing at Loch Scavaig, looking W.
E, slopes of Sgùrr nan Eag; G, An Garbh-choire; D, Sgùrr Dubh; A, Allt a' Chaoich; p, peridotite; g, gabbro.

Fig. 66. W. side of Strathaird Peninsula
B, Ben Cleat; E, Elgol; C, caves; i, Inferior Oolite sandstone; e, Great Estuarine Series (sandstones and shales with oyster beds); o, Oxfordian shales and sandstones; b, basalt lavas; d, dolerite sill.

opposite side of the ridge. The slabs of glaciated gabbro on the shore and in the lower slopes are an impressive sight, and another memorial of the ice-age is seen in the numerous 'perched blocks' on the slopes of the mountains.

Eastward the basalts reappear above sea-level and then the Torridon Sandstone, which floors the valley running inland from Camasunary, but is then cut off by an important fault. The Strathaird peninsula, which

94

makes the eastern side of Loch Scavaig, is of Jurassic rocks, capped by the basalt lavas on the higher ground (fig. 66). After rounding the southern headland, with its numerous caves, we face the wide opening leading to Lochs Slapin and Eishort (fig. 67). All the eastern coast of Strathaird is made by Jurassic rocks (mostly the Inferior Oolite sandstones), but Blaven and the Broadford group of Red Hills are seen in the distance. The promontory dividing Loch Slapin from Loch Eishort is of Lias, capped by a composite sill of granophyre bordered by basalt (see p. 79). S. of Loch Eishort is a tract of older rocks, overthrust and faulted. They are mainly Torridonian with some Cambrian, the white quartzite hills E. of Ord making a conspicuous feature. This is the character of the country as far as the Point of Sleat, after passing which we enter the Sound of Sleat, already described (p. 58).

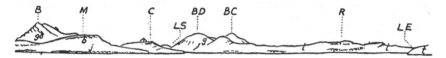

Fig. 67. Looking to Loch Slapin

B, Blaven; M, Ben Meabost; C, An Carnach, with the summit of Beinn na Crò behind; LS, entrance of Loch Slapin; BD, Beinn Dearg Mhòr; BC, Beinn na Caillich; R, Rudha Suisnish; LE, entrance of Loch Eishort; t, Torridon Sandstone; l, Lias; j, Jurassic; b, basalt; gb, gabbro; g, granite.

XIV. KYLE OF LOCHALSH TO CAPE WRATH AND DURNESS

(a) KYLE OF LOCHALSH TO GRUINARD BAY

We shall follow now the coast-wise route northward, although this is not covered by any one regular service of steamers. Clearing the Kyle, we open a view of Raasay and some smaller isles, with a distant glimpse of Skye (fig. 68). Passing the wide mouth of Loch Carron and the Torridonian islands of Longay and Crowlin, the route lies through the Inner

Sound. On the right the Applecross district consists wholly of Torridon Sandstone, excepting only a small outlier of Trias and Lower Lias upon which the village stands. To the west Raasay is in full view, presenting a range of cliffs made by members of the Jurassic system; but, where the height declines, a considerable fault brings on the Torridon Sandstone. A little farther on the Lewisian gneisses emerge, and, with characteristic scenery, make the low-lying northern end of Raasay and the neighbouring island of South Rona.

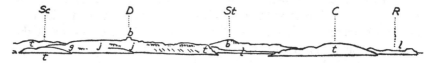

Fig. 68. Leaving the Kyle

Sc, Scalpay, with Longay in front; *D*, Dun Caan, the highest point of Raasay; *St*, Storr, in Skye, seen over the low northern part of Raasay; *C*, Crowlin; *R*, South Rona; *l*, Lewisian gneiss; *t*, Torridon Sandstone; *j*, Jurassic; *b*, Tertiary basalts; *g*, granite.

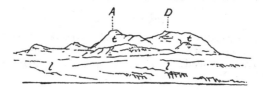

Fig. 69. Seen from Lower Loch Torridon

A, Beinn Alligin; *D*, Beinn Dearg; *l*, Lewisian gneisses; *t*, Torridon Sandstone.

On the mainland Torridon Sandstone continues, and indeed alternations of Torridonian and Lewisian rocks, with some pre-Lewisian, occupy all the rest of the west coast. The first place of interest is Loch Torridon, with its characteristic scenery. The entrance, where it is narrowest, is flanked on both sides by Lewisian rocks, but the sandstone is seen beyond (fig. 69), and a cruise into the upper loch reveals a typical Torridonian topography. The mountains are more individual than most of those carved out of the Highland crystalline schists, and they present unusually bold outlines. Some of the peaks show a capping of white Cambrian quartzite. The Lewisian inliers, which emerge from beneath the sandstone in the tract surrounding Loch Shieldaig, may be examined at Shieldaig village, a regular landing-place. The rocks are mainly micaceous gneisses, rich in quartz and with pegmatite streaks. Some bands are garnetiferous; others, less acid, carry hornblende as well as biotite. Traversing the gneisses in a N.W.–S.E. direction are numerous bands of amphibolite, the metamorphosed representatives of basic dykes.

The rather dreary coast north of Loch Torridon calls for no comment; but Gairloch offers numerous points of interest to the geologist. Approaching the pier the reddish Torridon Sandstone still continues on both sides, but the eastern part of the loch is in a tract of the older rocks. The Lewisian gneisses show some variety, the commonest being a granite-gneiss with light and dark micas, and there are broad belts of amphibolite and hornblende-schist running N.W.–S.E. Of most interest, however, is an area of the pre-Lewisian rocks of sedimentary origin, now represented by schists of various types. These may be seen to advantage in Kerrysdale, between Kerrysdale House and the Kerry Falls. The prevalent types are mica-schists and

Fig. 70. *Approaching Loch Maree*

S, Slioch; l, Lewisian gneiss; a, massive amphibolites in the Lewisian; t, Torridon Sandstone.

hornblende-biotite-schists, often garnetiferous; but there are in addition magnetitite-quartz-schists, calc-mica-schists, actinolite-schists, and marbles. Especially good exposures occur along the old road on the hillside.

The road on to Loch Maree passes everywhere over pre-Lewisian and Lewisian rocks, except for a patch of basal Torridonian as the loch is approached (fig. 70).

The Loch Maree hotel is on Lewisian gneiss, but just beyond the Torridon Sandstone comes on. The islands are also of the sandstone; but on the opposite side of the loch, which conceals an important line of faulting, the older rocks reappear. In the east is seen the bold outline of Slioch made by Torridon Sandstone (fig. 70).

Gneisses and amphibolites are well exposed near Gairloch pier and on the headland to the west. There are gneisses with two micas and others with hornblende and epidote. North of the headland a tract of blown sand makes the golf-course and beyond, at the hotel, the basal conglomerate of the Torridonian is finely exposed. The Poolewe road crosses first a belt of amphibolite and then a tract of gneisses like those last mentioned.

[From Gairloch by sea around the sandstone cliffs of Rudh' Ré Torridonian rocks continue past the lighthouse into Loch Ewe. After passing the Isle of Ewe Lewisian gneiss forms the S.E. coast, being

Fig. 71. *Gruinard Bay*

G, Beinn Ghobhlaich; *l*, Lewisian gneiss; *t*, Torridon Sandstone.

brought against the Torridonian on the lake-strewn lower ground by a continuation of the Loch Maree fault. On the S.E. side of the Isle of Ewe Trias sandstone rests unconformably upon the Torridonian. The outcrop is extended on the adjacent mainland from Aultbea up a hollow crossed by crescentic moraines.

After rounding Greenstone Point with its Torridonian cliffs, the Mesozoic outcrop is encountered again on the S. coast of Gruinard Bay, where exposures occur both of Trias sandstone and overlying Lias limestone. The head of the bay is bounded by gneiss, but Gruinard Island and the mainland on either side are made mainly by the Torridonian. From here the view to the E. also shows Beinn Ghobhlaich, a typical sandstone mountain (fig. 71).]

(b) GRUINARD BAY TO DURNESS

[Northwards of Gruinard Bay, from the mouth of squally Little Loch Broom, the 'Loch of a Hundred Winds', the mountains farther inland formed of the red Torridonian Sandstone may be seen again, with precipitous Beinn Ghobhlaich N. of the loch and the varied outlines of An Teallach farther S. with its white capping of Cambrian quartzite.

Cailleach Head is made by the stratigraphically highest Torridonian beds, which have been brought down against lower strata by a N.W. fault that crosses the promontory. At the extreme point a band of shale, the characteristic associate of the light coloured red sandstones of the Upper

Fig. 72. Unconformity seen at Ullapool, just E. of the village
t, Torridon Sandstone; q, Cambrian quartzite; r, raised beach.

Torridonian, forms a conspicuous feature along the strike. A short way beyond, a second promontory is made by Lewisian gneiss, the outcrop of which extends uphill to the little summit, Cnoc Sgoraig, with Torridonian beds on either side. This is one of many examples of the uneven surface of gneiss upon and against which the Torridonian sediments were deposited. The gneiss outcrop is continued to the N. through two small islands to Horse Island and the adjoining mainland where the Cailleach Head fault reappears. Seawards, beyond the fault, lie the Summer Isles, chiefly Upper Torridonian.

Loch Broom cuts inland across the Torridon Sandstone to the main outcrop of the Cambrian and through the overthrusts which accompany this belt of rocks to the limiting Moine Thrust and the overriding Moine schists beyond. Half-way up the loch, on the hillside above Ullapool, the Cambrian quartzite rests at an angle unconformably upon the Torridonian, with a basal escarpment consisting of a series of white crags (fig. 72). Some higher Cambrian beds are succeeded eastward by two thrust masses, the lower of Torridonian and Cambrian strata, the upper of Lewisian gneiss. The latter is surmounted along the Moine Thrust-plane by the Moine schists which form all the mountains farther E.

The Summer Isles and the promontory, Rhu More, to the N. are topographically a series of ridges and hollows which run N.N.E. parallel

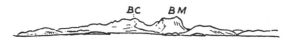

Fig. 73. Coigach, seen from the S.W.

BC, Beinn nan Caorach; BM, Ben More Coigach. All is of Torridon Sandstone.

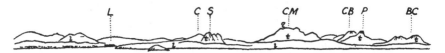

Fig. 74. Approaching Lochinver: Outlines of the Assynt Mountains

L, Lochinver; C, Canisp; S, Suilven; CM, Cùl Mòr; CB, Cùl Beag; P, Stac Polly (An Stac); BC, Ben More Coigach; l, Lewisian gneiss; t, Torridon Sandstone; q, Cambrian quartzite.

to the strike of the beds. This kind of scenery is typical of the steeply dipping Upper Torridonian sandstones and is a contrast to the imposing mountains to the E. which are made by the flat-lying and more wide-spread lower beds (fig. 73).

After passing Rhu Coigach, the approach to Lochinver shows a wide panorama of the Assynt Mountains (fig. 74). On a floor of Lewisian gneiss are set the characteristic giant remnants of the denuded Torridon Sandstone, a series of mountains which display an endless variety of cliff, buttress, peak and corrie. Their warmer hues are in contrast to the quartzite-capped summits of Canisp and Cùl Mòr (figs. 74, 75).]

Just above Lochinver, on the road to Inchnadamph, the Lewisian near Inver Bridge is mostly of granulitic hornblende-gneiss, often well banded. Pyroxene-gneiss, however, is much more typical of the Lochinver-Loch Broom area. About ¼ mile farther on is a conspicuous red dyke. It is an augite-oligoclase-porphyrite of a type which makes conspicuous sills in the country farther E., and is known as the Canisp porphyrite. About ⅓ mile farther along the road the gneiss is seen traversed by belts of shearing, the original coarse streaky structure being replaced by a new banding and foliation.

A drive to Inchnadamph affords numerous features of interest. The road runs at first over gneiss traversed by numerous basic dykes, now in

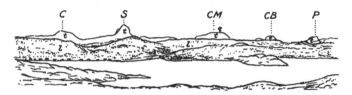

Fig. 75. Outlines of the Assynt Mountains, from Lochinver

C, Canisp; S, Suilven; CM, Cùl Mòr; CB, Cùl Beag; P, Stac Polly (An Stac); l, Lewisian gneiss; t, Torridon Sandstone; q, Cambrian quartzite.

the state of hornblende-schists or amphibolites. At Loch nan Eun, more than a mile from Lochinver, we cross one of the ultrabasic dykes of this district, and a little farther, at Brackloch, the road runs for some distance along another of the group. They are of picrite and trend slightly N. of W. The road now follows the River Inver to Loch Assynt and crosses many of the basic dykes running N.W. and one of picrite.

As Loch Assynt is approached, the precipitous face of Quinag, a characteristic piece of Torridon Sandstone architecture, is a striking sight. About half-way along the loch the sandstone comes down to the road, and its unconformable relation to the gneiss can be verified. A mile farther on it gives place to the Cambrian quartzite, first the gritty, false-bedded basal member and then the 'pipe-rock' with its worm-tubes.

Looking back we obtain a different view of Quinag, and may observe how the summit and all the eastern side is made by a dip slope of the quartzite resting unconformably upon the gently inclined Torridonian strata (fig. 76).

Following the ascending succession of the Cambrian, we come, after passing Skiag Bridge, to the shaly Fucoid Beds and the carious-weathering Serpulite Grit; but here we have entered the highly disturbed 'zone of imbrication' (see p. 6), which involves also the succeeding limestones (with sills of spessartite), seen about Ardvreck Castle and on to Inchnadamph.

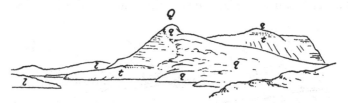

Fig. 76. Quinag from the S.E., with Loch Assynt

Q, Quinag; *l*, Lewisian gneiss; *t*, Torridon Sandstone; *q*, Cambrian quartzite.

[From the hillside behind Inchnadamph the view looking back shows much that has been already described as well as some new details. To the right, Quinag with its dip-slope of Cambrian quartzite lies beyond Loch Assynt and a hill topped by a monument which was erected as a memorial to geologists who unravelled the complicated Assynt country. To the left rises Canisp with its sills of porphyrite from behind a ridge made by Cambrian quartzite and acid sill-intrusions (fig. 77). Nearby is a crag of Cambrian limestone which belongs to the 'zone of overthrusts' (p. 6).

Inchnadamph is a centre well known to geologists from which to explore the classic geology of the district.[1] Here the 'zone of overthrusts' covers an exceptionally wide area which extends for 7 miles eastward to the limiting Moine Thrust. The belt narrows again some 6 miles to the

[1] A short guide is published by the Edinburgh Geological Society.

N., at Loch Glencoul, where the effects of the vast displacements may be appreciated from a distance. The best view-point, near Unapool House, may be reached either by road from Inchnadamph or by sea (fig. 78).

To the left, on the hillside across the loch, the unmoved rocks of the 'foreland region' are seen; but, unlike the many other views of it which

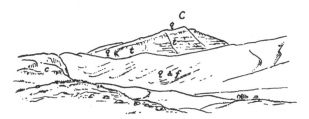

Fig. 77. Canisp from behind Inchnadamph

C, Canisp; t, Torridon Sandstone; q, Cambrian quartzite, with acid sills (f); c, Cambrian limestone.

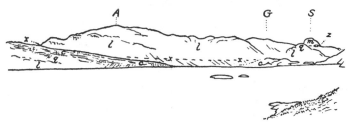

Fig. 78. Loch Glencoul, looking E. from Unapool

A, Beinn Aird da Loch; G, Glen Coul; S, Stack of Glencoul; l, Lewisian gneiss; q, Cambrian quartzite; c, Cambrian limestone, Serpulite Grit and Fucoid Beds; m, flaggy Moine schists; x, Glencoul Thrust-plane; z, Moine Thrust-plane.

have been described hitherto, there is a vast unconformity at the base of the Cambrian quartzite. The Torridonian has been completely removed by denudation and the Cambrian beds rest directly upon Lewisian gneiss. The 'sole' of the lowest displacement, the Glencoul Overthrust, along which Lewisian gneiss was pushed westwards across the Cambrian, may be followed by the eye to the head of the loch. Above this point, on the sky-line, is the Stack of Glencoul, made by Moine schists. These with the Moine Overthrust between are riding upon the Cambrian quartzite which, here again, rests unconformably directly upon the gneiss.

The coast-line between Lochinver and Cape Wrath, with its headlands, fiord-like lochs and island-fringes, repeats many features of the seaboard farther south. The scenery for many miles inland, however, is milder, though still rugged with much bare grey rock. The geological explanation is the predominance of the Lewisian gneiss as the country-rock. The mountain-making Torridon Sandstone is confined to the low headland

Fig. 79. *Scourie Bay and Handa Island*
H, Handa Island; l, Lewisian gneiss; t, Torridonian.

around the Point of Stoer, Handa Island N. of Scourie Bay (fig. 79), and a more extensive development N. of Loch Inchard towards Cape Wrath. The bold cliffs of that grim promontory are made by well-banded gneiss.

Eastward of Cape Wrath Torridonian again forms the coast, with a large capping of Cambrian quartzite on Scrivishven S. of Garve Island, the latter made by Cambrian limestone. The island belongs to a fault-block, the fault concerned crossing land at the entrance to the Kyle of Durness, where it brings the limestone against Lewisian gneiss inland, and again at the base of Faraid, the eastern point of the Kyle. The displacement due to the fault must be enormous. It brings rocks belonging to the 'zone of overthrusts', which make the Faraid headland, against the unmoved Cambrian of the 'foreland', 6 miles westward of the main outcrop of the zone on the E. side of Loch Eriboll. It is a

6-mile displacement in the horizontal direction, and the fault is presumably a wrench or tear fault.

The upper reach of the Kyle of Durness is a sandy flat at low tide traversed by the meanders of the River Dionard. To the west is mainly bare Lewisian gneiss, varied by two broad outcrops of Torridon Sandstone and Cambrian quartzite. To the east, around Durness village and southward to the head of the Kyle, is a grassy tract of cream-coloured rock, the type locality for the Cambrian limestone. Here the formation attains a thickness which exceeds that encountered in any other part of the North-West Highlands.]

XV. THE OUTER HEBRIDES AND THE SHIANT ISLES

(a) THE OUTER HEBRIDES

[The islands of the Outer Hebrides, collectively termed the 'Long Island', are Lewis and Harris to the N.E., then North and South Uist separated by smaller Benbecula, and Barra with its attendant islands at the S.W. end of the entire group. They present much diversity of scenery, but perhaps most characteristic is an intimate interlacing of land and water, both salt and fresh. Fiord-like sea-lochs, often indescribably intricate, are especially numerous along the eastern seaboard. In association there is an abundance of small promontories, islands and reefs. Unlike Skye and the mainland across the Minch, raised beaches are absent. The tract indeed is one of subsidence and not uplift, and direct evidence is supplied in various places by the occurrence of 'drowned' peat extending down to low-water mark or even lower. The fiords themselves are 'drowned' valleys, and sometimes give place inland to brackish-water lochs and, finally, to fresh-water lakes.

A large part of the islands though rock-covered and rugged is low lying, especially in the central and northern half of Lewis, over Benbecula, and westwards of an eastern border of mountains in South Uist. To the W., as in South Uist, Benbecula and elsewhere, is typical machair land, with a sandy coast fringed with dunes from where shell sand is swept inland and has served to fertilize the ubiquitous peaty soil. Harris especially is mountainous, Clisham in North Harris rising to 2622 ft.

The islands are made essentially of Lewisian gneiss, with which is associated a large mass of pre-Lewisian paragneiss in the S.W. part of South Harris. The strike of the foliation both of the paragneiss and Lewisian orthogneiss is dominantly N.W.–S.E., as on the mainland. The paragneisses include crystalline limestones, quartzose rocks, graphite-schists, and garnetiferous quartz-schists, sillimanite-gneisses and kyanite-gneisses. There is therefore a resemblance to the altered calcareous and graphitic pre-Lewisian sediments of the mainland and the Inner Hebrides (pp. 61, 65). The most widespread variety of orthogneiss is a banded grey or pinkish granitic rock with biotite or hornblende, rarely with pyroxene. Hornblende-gneiss, often richly garnetiferous, is a frequent associate, in lenticular and pillow-like masses. More acid pink granite-gneiss with two micas, dark and light, is also fairly common and sometimes, as in S.W. Lewis where it is intrusive in the grey gneiss, it forms large masses. It tends to make country with smooth outlines. Pegmatite veins are ubiquitous. A rather exceptional rock called anorthosite-gneiss, consisting mainly of labradorite felspar, forms the hill of Roneval in South Harris where it is associated with a mass of gabbro-diorite.

Intrusive in the gneisses are elongate masses, sheets and dykes composed of various kinds of rock. Many are late-Lewisian in age and occur often in broad and relatively short intrusive bodies. They are composed of peridotite, norite, biotite- and hornblende-bearing basic rocks, etc. Many more are typical minor intrusions, mainly dykes. Lamprophyres belonging to the Caledonian series, camptonites probably of Permian age, and Tertiary N.W. dykes of olivine-dolerite, crinanite and quartz-dolerite, are all represented.

Later than the Lewisian, too, are certain unaltered bedded sediments composed of materials derived from the gneiss. These are the Stornoway

Beds, a series of reddish-brown conglomerates and sandstones which extend for some distance N. and E. of the town. They belong to some continental formation laid down by the action of torrents in a region of youthful topography. Their age is unknown. They might be assigned to any one of the three great continental formations of Scotland, the Torridonian, the Old Red Sandstone, or the Trias.

Of earlier date than the Stornoway Beds, in fact probably pre-Torridonian, is a structural feature of major significance. This is a thrust-plane gently inclined towards the S.E. which extends along practically the entire S.E. side of the Long Island, from a point a few miles S. of the Butt of Lewis to Barra and the smaller islands beyond. The strip of gneiss E. of the thrust consists of highly crushed and sheared rocks, and the thrust-plane itself is often marked by a development of flinty-crush rock up to 100 ft. in thickness, a dark heavy vitreous material. The gneiss along the dislocation was rendered semi-molten or even liquid by the intense stresses set up through the movement of the overlying thrust-gneisses. The movement was towards the N.W. Often anastomosing irregular veins of flinty-crush rock are also encountered in the crushed gneisses, and even in the underlying and otherwise unaffected varieties.

The ports of call for steamers are all situated along the S.E. coast. They are: Stornoway in Lewis, Tarbert and Rodil in North and South Harris, Lochmaddy in North Uist, Lochboisdale in South Uist, and Castlebay in Barra. Only a brief mention of the rocks at these various places will be made.

At Stornoway, the Stornoway Beds rest unconformably on a floor of crushed gneisses on the W. side of the Eye Peninsula and E. of the town.

At Tarbert, W. of the thrust, reddish-coloured Lewisian gneiss forms the hills to the N. Rodil is a centre from which to see the pre-Lewisian paragneisses, as well as the Lewisian anorthosite-gneiss and gabbro-diorite which make the hilly belt to the N.

Astride and S. of the entrance to Loch Maddy, three rocky islets are remnants of an inclined sill of olivine-dolerite. Within the entrance the thrust-plane extends N. and S. around the hills on either side; the remainder of this loch with its many arms is cut in the uncrushed orthogneiss.

The base of the crushed gneiss again extends obliquely up a hillside N. of the entrance to Loch Boisdale, and along it flinty-crush rock is developed on a great scale. Near the hotel, in addition to the prevalent biotite-gneiss, some hornblende-pyroxene-gneiss is exposed.

In Castle Bay flinty-crush rock forms the rocky islet on which Kiesimul Castle stands, and is well exposed on the foreshore immediately E. of the pier, where the rock looks at first sight like an intrusion of basalt many feet in thickness. Along the promontory to the W. a few basic dykes are exposed.]

(b) THE SHIANT ISLES

These islands attract the notice of a traveller to Stornoway or Tarbert; but, since they are uninhabited, a would-be visitor must make his own arrangements, and a brief notice will suffice in this place. Although lying

Fig. 80. The Shiant Isles, seen from the N.W.

M, Eilean Mhuire; G, Garbh Eilean; T, Eilean an Tighe; GB, Galta Beag; l, Lias shales; c, crinanite of main sill; t, talus concealing shales; x, lower sill of Garbh Eilean; z, lower sill of Eilean Mhuire.

near to Lewis, their geological relations are with Skye, and they may be regarded as the most northerly of the Inner Hebrides, in which Tertiary igneous rocks make the leading features.

The group consists of three principal islands with a string of rocky islets stretching away to the W. (fig. 80). The three larger members partly enclose a sheltered lagoon, which swarms with sea-birds of various kinds, whose nests line the surrounding cliffs. The two westerly islands, viz. Garbh Eilean and Eilean an Tighe, lie close together, and are indeed united by a shingle beach. They are made essentially by one great intrusive sill, more than 500 ft. thick, with no visible top. It dips to S. and W. so that the northern and eastern sides present bold cliffs, and the surface declines to S. and W. along what is roughly a dip slope (fig. 80). The northern face of Garbh Eilean is a nearly vertical cliff showing a section of the great sill with regular columnar jointing on a large scale, the columns being 5 or 6 ft. in diameter. At the base Upper Lias shales emerge from below, but these are mostly concealed, like the sill itself, by a talus of large blocks. Below these shales is a lower sill, which makes the rocks at tide-marks, and extends eastward as a low promontory pierced by a curious natural tunnel (fig. 81).

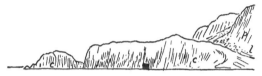

Fig. 81. The N.E. point of Garbh Eilean, Shiant Isles

The low cliff in the foreground, pierced by a natural arch or tunnel, is made by the lower crinanite sill (*c*). Overlying this are Liassic shales (*l*), largely hidden by a talus of blocks from the picrite (*p*) which forms the base of the main sill.

The sills are of crinanite. The great one is a very clear illustration of 'gravitative differentiation'. Owing to the sinking of the relatively heavy olivine crystals, the proportion of that mineral increases downward, until at the bottom the basic rock has become ultrabasic (picrite). This latter type is seen at the base of the cliff on the E. side of Garbh Eilean: farther N. it is mostly hidden by talus.

The small islets, Galta Beag, Galta Mòr, etc., are probably relics of the great sill. Eilean Mhuire, lying to the E., is different. It consists of two crinanite sills separated by shales, but the lower sill is of a somewhat composite nature. Its middle part, as seen on the south-eastern promontory, is of analcime-syenite, and there are coarse segregations rich in analcime and containing nepheline.

EXPLANATION OF TOPOGRAPHICAL MAPS V–VIII

The numerals 1–81 are placed approximately at, or near, the view-points of the similarly numbered figures in the text.

The sea routes indicated by various types of line are those described in the text. The accompanying Roman numerals (II, etc.), sometimes with letters (*a*, *b*, etc.), refer to the chapter, or section of chapter, which deals with the route concerned.

Abbreviations: B. (suffix), Bay; *B.* (prefix) and Bh. (suffix), Ben, Beinn or Bheinn; Cas., Castle; G. and Gl., Glen; Hd., Headland; Ho., House; I., Isle or Island; Is., Isles or Islands; L., Loch; N., North; Pt., Point.

ERRATA

Beinn Gobhlaich=Beinn Ghobhlaich
Cul Bheag=Cul Beag
Farrid=Faraid
Scrishven=Scrivishven or Sgribhir-bheinn

Map V

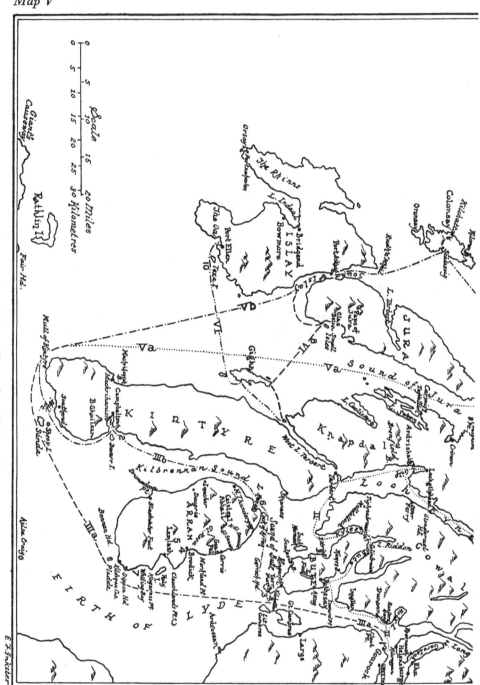

Topographical Map: From the Clyde to Jura

Map VI

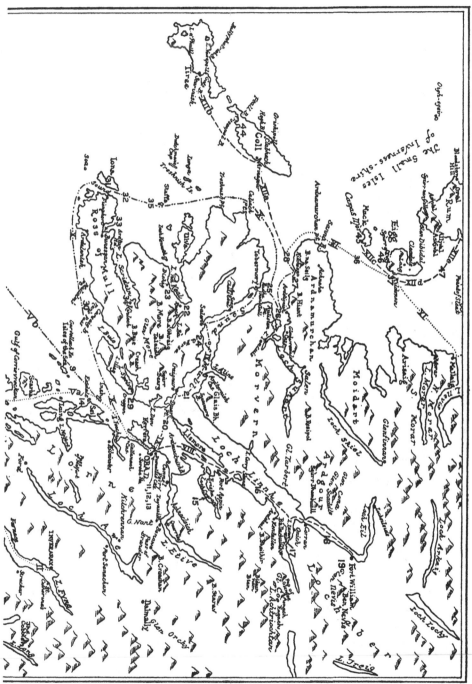

Topographical Map: From Colonsay to Rum

Map VII

Topographical Map: From Rum to Gairloch

E. F. Baxter

Map VIII

Topographical Map: From Loch Ewe to Cape Wrath

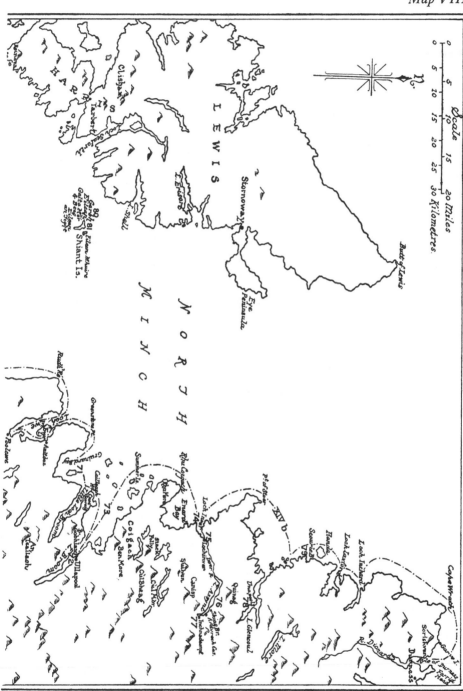

GLOSSARY OF GEOLOGICAL TERMS, MINERALS, AND ROCKS[1]

Acid rocks. Igneous rocks with a high content of silica (over 65 per cent.).

Actinolite. A green mineral of the amphibole group occurring in elongate, very narrow crystals; characteristic of certain metamorphic rocks especially of altered dolomite.

Aegirine. A dark green soda-rich mineral belonging to the pyroxene group; characteristic of many alkali-rich igneous rocks.

Agglomerate. Mixture of blocks, fragments and finer debris thrown up by a volcano.

Albite. A variety of plagioclase felspar, the richest in silica and in soda.

Allivalite. A coarsely crystallized ultrabasic igneous rock, consisting essentially of basic plagioclase felspar and olivine.

Amphibole. A group mainly of ferromagnesian minerals, often occurring in elongate crystals; especially characteristic of intermediate igneous and of certain metamorphic rocks.

Amphibolite. A metamorphic rock largely composed of amphibole and plagioclase.

Amygdale (Gk. *amugdalon* = almond). Small cavity (vesicle, often almond-shaped) formed in great numbers mainly in the upper parts of lava-flows, also in dykes, as a result of expansion of contained gases (mainly steam) while the rock was molten, and subsequently filled in with zeolites and other hydrous minerals, sometimes with quartz or calcite.

Amygdaloidal. Applied to rocks containing amygdales.

Analcime. A hydrous milk-white mineral allied to the zeolites, rich in soda.

Andalusite. A brownish red mineral, composed of silicate of alumina, occurring in stumpy crystals square in cross-section; developed in fairly highly metamorphosed argillaceous rocks.

Andesite. A rather dark fine-grained or compact intermediate igneous rock; as lava-flows and sometimes dykes and other minor intrusions.

Anorthite. Variety of plagioclase felspar, poorest in silica and richest in lime. Colour often grey with slightly greasy lustre.

Anticline. An elongate convex fold in bedded rocks which dip in directions away from the axial plane of the fold, like an inverted V.

Arenaceous (L. *arena* = sand). Applied to sedimentary rocks which were formed largely of sand.

---

[1] Rock names are modified to indicate varieties by prefixing the name of the conspicuous mineral or minerals present in addition to the essential constituents; e.g. olivine-basalt.

117

Argillaceous (L. *argilla*=white clay). Applied to sedimentary rocks, of any colour, which were formed largely of clay or mud.

Arkose. A sedimentary rock composed mainly of quartz and felspar; derived from disintegrated granite.

Augite. A common mineral of the pyroxene group, especially characteristic of basic and many ultrabasic igneous rocks.

Barytes. A heavy usually pink mineral with good cleavage and vitreous lustre, composed of barium sulphate; usually in veins.

Basalt. A dark fine-grained basic igneous rock, composed essentially of labradorite plagioclase felspar and augite; as lava-flows and as dykes and other minor intrusions.

Basic rocks. Igneous rocks with a relatively low content of silica (45–55 per cent.).

Bed. As applied to sedimentary rocks, an individual layer of sediment (*adj.* bedded). Very thin beds are termed *laminae* (*adj.* laminated).

Biotite. A dark usually brown-coloured variety of mica, rich in iron and magnesia; characteristic of granite and many schists and gneisses.

Bostonite. A pale grey fine or medium grained alkaline intermediate igneous rock; usually in dykes and sills.

Boulder-clay. A stiff clay with stones of various sizes, often polished and striated; formed beneath an ice-sheet as a ground-moraine.

Breccia. A rock either of sedimentary origin or due to volcanic explosion, composed of angular rock-fragments. Also formed by crushing (crush-breccia).

Calcite. A white translucent or transparent mineral with well-developed cross cleavages; composed of carbonate of lime. In distinction to felspar, easily scratched with a knife.

Camptonite. A fine-grained basic or ultrabasic igneous rock allied to basalt, but containing hornblende and analcime; little spots (ocelli) of pink analcime penetrated by small dark laths of hornblende are scattered through the rock, which often weathers with a finely knobby surface; as dykes.

Carious. Applied to weathered rock-surfaces, often of limestone, with a coarsely pitted appearance.

Cement-stone. An argillaceous limestone very fine in grain; named from the fact that such rocks have been used for making cement.

Chalcedony. Amorphous and minutely crystalline silica, sometimes finely banded and usually occurring in vesicles of lavas.

Chalk. A white extremely pure limestone constituted mainly from the minute calcareous shells of marine Protozoa (Foraminifera).

Chamosite. A dark green mineral composed of ferrous aluminium silicate, the essential iron constituent of a Jurassic (Upper Lias) bed of ironstone.

Chert. Impure flint, breaking with a flat fracture, in nodules and bands usually associated with shales, or with limestones.

Chlorite. A dark green soft flaky hydrous ferromagnesian mineral; a constituent of lowly metamorphosed pelitic rocks; a common alteration product of pyroxene and other ferromagnesian minerals in igneous rocks; also occurs filling vesicles of lavas, etc.

Columnar jointing. Joints typically joining in groups of three and forming six-sided (hexagonal) columns in lavas and sills.

Composite. Applied to minor intrusions (dykes and sills) which consist of marginal and central portions differing in composition and not chilled against one another. It is concluded that the central portion, which is the later intrusion of the two, penetrated the earlier injection before the latter had cooled completely.

Cone-sheet. A member of an arcuate or ring-shaped belt of inclined intrusive sheets which, individually, are inclined downwards and inwards towards the centre of the belt. Cone-sheets occur in great numbers around several of the Tertiary igneous centres of the Inner Hebrides where they are usually associated with ring-dykes.

Conglomerate. A sedimentary rock composed mainly of pebbles.

Cordierite. A mineral composed of silicate of alumina, magnesia and iron; occurring in highly metamorphosed sedimentary rocks and imparting a greasy appearance.

Craignurite. A finely crystallized medium-grey intermediate igneous rock containing a good proportion of augite in small elongate crystals in addition to felspar; in Tertiary minor intrusions.

Crinanite. Allied to olivine-dolerite, but containing analcime and with basic plagioclase felspar; in Tertiary dykes and sills.

Current-bedding. A common kind of false-bedding due to water or wind currents, with bedding-planes set at an angle to the true bedding; they mark surfaces of deposition which were inclined to the horizontal, as in the case of sand-banks and sand-dunes.

Cyanite. A white or blue mineral forming long blade-like crystals, composed of silicate of alumina and occurring in highly metamorphosed argillaceous rocks.

Diopside. A lime-rich ferromagnesian mineral of the pyroxene group, usually medium or dark green in colour; occurs both in igneous and metamorphic rocks.

Diorite. A coarsely crystallized intermediate intrusive igneous rock, composed essentially of intermediate plagioclase felspar and hornblende; usually in major (plutonic) intrusions.

Dip. Inclination of bedding planes of sedimentary rocks.

Dolerite. A dark basic igneous rock of medium grain composed essentially of labradorite plagioclase felspar and augite; in dykes, sills, cone-sheets, etc.

Dolomite. A rock (or a mineral) composed of carbonates of lime and magnesia. As a rock, usually originated as a limestone which was altered by the infiltration of magnesian solutions.

Drift. Unconsolidated superficial deposits, including those left by the Pleistocene ice.

Dyke. A highly inclined or vertical intrusive sheet of igneous rock, usually of great length and only a few feet or perhaps yards in width.

Dyke-swarm. An assemblage of dykes which is coincident usually with a plutonic complex of the same general age. The term includes linear assemblages of parallel dykes (linear swarm), and groups of dykes radially disposed to a plutonic complex (radial swarm).

Eclogite. A basic igneous rock of medium or fine grain, containing garnets; usually of metamorphic origin, associated with metamorphic rocks which have been subjected to pressure.

Epidiorite. Metamorphosed basic igneous rock, usually originally a dolerite or gabbro, the most striking change being the conversion of augite into hornblende.

Epidote. A bright green mineral, composed of silicate of alumina, ferrous iron and lime; of common occurrence in altered rocks in which lime in particular has been made available.

Eucrite. A coarsely crystallized basic intrusive igneous rock, differing from gabbro in containing *basic* plagioclase felspar; in major (plutonic) intrusions.

False-bedding. Irregular bedding, the planes of which lie at an angle to the true bedding. *See* Current-bedding.

Fault. A steeply inclined plane of fracture which is accompanied by displacement of the rocks on one side of the fracture relatively to those on the other. In most cases, where the rocks have been in a state of tension, the displacement is downwards (normal fault), but in regions of compression (mountain belts) may be upwards (reversed fault), or it may be even sideways (wrench or tear fault). In the last-mentioned case the fracture itself is typically vertical. *See also* Hade.

Fayalite. An iron-rich variety of olivine.

Felsite. A fine-grained or compact pale, often pink or red, acid igneous rock, composed mainly of orthoclase felspar and quartz with a small amount of biotite or hornblende; in minor intrusions.

Felspar. A white, grey or pink mineral with good cleavage, giving plane mirror-like faces on fracture, a constituent of almost all igneous rocks; composed of silicate of alumina, with potash (*orthoclase* and *microcline*, often pink or red), or with soda and lime in varying proportions (*plagioclase*: *see* Albite, Oligoclase, Labradorite, Anorthite).

Ferromagnesian. Applied to minerals composed, partly or wholly, of silicates of iron and magnesia.

Flint. A dark grey rock composed of microcrystalline silica (Chalcedony), translucent and with a curved (conchoidal) fracture; occurs as nodules in the chalk.

Flinty crush-rock. A black semi-glassy-looking rock formed by very intense crushing of gneisses and other rocks; occurs as veins ramifying through the parent rock, or as larger bodies along thrusts (e.g. Outer Hebrides).

Fluvio-glacial. Applied to deposits laid down by melt-waters within or issuing from a glacier or ice-sheet.

Foliation (L. *folium* = a leaf). A parallel arrangement of minerals formed in metamorphic rocks.

Forsterite. A magnesia-rich variety of olivine.

Gabbro. A coarse-grained basic igneous rock composed mainly of labradorite felspar and augite, often with olivine (olivine-gabbro); in major (plutonic) intrusions.

Garnet. A group of minerals variously composed of silicates of lime, iron, alumina, etc., in equidimensional many-sided crystals (dodecahedra). A dark red variety (almandine) is common in the Highlands.

Glauconite. A finely crystalline green mineral, a hydrous silicate of iron and potash; largely constituting glauconitic oozes of the present sea-floor; occurs in abundant small grains in the Cretaceous greensand.

Gneiss. A class of coarsely crystallized and usually well-banded metamorphic rocks. *See* Orthogneiss, Paragneiss.

Granite. A coarse-grained acid igneous rock composed mainly of quartz, orthoclase felspar and mica or, less commonly, hornblende; in major (plutonic) intrusions.

Granodiorite. A coarse-grained igneous rock intermediate in composition between granite and quartz-diorite, and containing much acid plagioclase felspar (oligoclase); in major (plutonic) intrusions, Caledonian in age.

Granophyre. A medium-grained acid igneous rock composed mainly of quartz, orthoclase felspar and a ferromagnesian mineral (biotite, hornblende or augite). It differs from granite mainly in the fact that the quartz and felspar have intergrown intimately with one another on a minute scale; typical plutonic or hypabyssal acid intrusive rock of the Tertiary period.

Granulite. A metamorphic rock composed of roughly equidimensional (not elongate) crystals, usually quartz and felspar.

Hade. The inclination of a fault, sometimes applied also to the inclination of a dyke. In the case of a normal fault, the hade is towards the displaced (downthrow) side, in a reversed fault towards the displaced (upthrow) side.

Hanging valley. A tributary valley which enters a main valley at a point well above the bottom of the main valley. The main valley has been deepened by a glacier, leaving the tributary valley 'hanging'.

Hornblende. A common mineral of the amphibole group, especially characteristic of intermediate igneous rocks.

Hornblendite. A basic igneous rock, usually coarse-grained, composed essentially of hornblende together with some felspar; usually in minor intrusions.

Hornfels. A fine-grained or compact rock, crystallized or recrystallized under the action of heat derived from an adjacent igneous intrusion.

Hypabyssal. Applied to intrusive igneous rocks which occur in small masses (minor intrusions) and have crystallized fairly rapidly. The crystal nature is in general medium grained, with crystals discernible without use of lens.

Hypersthene. A mineral of the pyroxene group, composed of silicate of iron and magnesia; frequently occurring in basic igneous rocks and crystallizing from magmas at higher temperatures than augite; also a product of thermal metamorphism (e.g. in hornfels).

Igneous. Applied to rocks which have crystallized from a molten state. Such rocks occur either as *lavas*, etc. (*volcanic* rocks) or as *intrusions*, either minor intrusions such as dykes and sills (*hypabyssal* rocks) or major intrusions (*plutonic* rocks).

Inclined Sheet. *See* Cone-sheet.

Intermediate rocks. Igneous rocks intermediate in silica content (55 to 65 per cent.) between acid and basic rocks.

Intrusion. A body of magma (liquid rock) which was injected into solid rocks and which thereafter solidified.

Isoclinal Fold. An acute fold in bedded rocks the axial plane of which is inclined, both limbs being inclined in the same direction.

Kentallenite. A medium or somewhat coarse grained basic igneous rock remarkable in that it contains biotite and orthoclase felspar; in major intrusions of Caledonian age.

Kersantite. *See* Lamprophyre.

Labradorite. A variety of plagioclase felspar intermediate in composition between albite and anorthite; a constituent of basic igneous rocks.

Laccolite (Gk. *lakkos* = a reservoir). A major intrusion in shape like a lens with a flat base and dome-like top.

Lamina. *See under* Bed.

Lamprophyre. A group of fine-grained basic igneous rocks made up of well-formed crystals of orthoclase (O) or plagioclase (P) felspar, combined with biotite (B) or hornblende (H). Thus four main varieties are recognized: Minette (O + B), Kersantite (P + B), Vogesite (O + H), Spessartite (P + H). Occurring in dykes and sheets of Caledonian age.

Magnetite. A black magnetic iron-ore, composed of ferric and ferrous oxide; occurring in small quantities in most igneous rocks.

Marble. Crystalline limestone, a metamorphic rock in the geological sense.

Marl. A soft unbedded or poorly bedded, characteristically limy, argillaceous rock.

Marscoite. A hybrid igneous rock formed by the intermixture of gabbro and granite, one of which, while in the molten state, attacked the other.

Metamorphism. A subject concerned with the more or less profound changes in mineralogy and texture set up in rocks in response to changes of temperature, stress, etc. (*adj.* metamorphic, *vb.* metamorphose). Such changes may be due to adjacent igneous intrusions (*contact-metamorphism*) or to stress (*dynamic action*) and rise of temperature in the interior of mountain-chains (*regional metamorphism*).

Microgranite. Fine-grained granite.

Monchiquite. A fine-grained ultrabasic rock allied to Camptonite but without felspar and usually rich in olivine. An olivine-free variety which is termed *ouachitite* is characterised by abundance of biotite.

Moraine. An accumulation of stones and sandy clay which was formed along the sides and, more extensively, along the front of a glacier; often in a series of mounds (*hummocky drift*). For Ground-moraine *see* Boulder-clay.

Mugearite. A fine-grained dark basic to intermediate igneous rock with a closely-set platy structure due to flow-movement of the partially crystallized magma, composed essentially of acid plagioclase felspar (oligoclase) and ferromagnesian minerals; usually in lava-flows, sometimes in minor intrusions.

Multiple Dyke. A group of dykes, with chilled margins, running side by side or, more usually, with one splitting another.

Muscovite. A white or greyish variety of mica rich in potash, occurring in some granites and orthogneisses, and in many pegmatite veins and schists.

Mylonite. A fine-grained very streaky rock, formed by intense shearing along thrusts.

Neck. The throat of a volcano, now exposed by erosion, infilled with solidified magma (*plug*) and/or with agglomerate.

Nepheline. A mineral similar in composition to albite plagioclase felspar, but with less silica; in soda-rich igneous rocks.

Norite. A coarse-grained basic igneous rock like a gabbro but with hypersthene instead of augite; in major (plutonic) intrusions.

Oligoclase. An acid, usually white, plagioclase felspar, with more soda than lime; an essential constituent of granodiorite, mugearite, etc.

Olivine. A ferromagnesian mineral in diamond-shaped crystals, green and semi-translucent when fresh, but usually altered to a bright red product; a common constituent of basic and ultrabasic rocks.

Oolitic (Gk. *oon* = egg, *lithos* = stone). Applied to a sedimentary rock, usually limestone, composed of spherical grains (oolites) resembling fish-roe, formed by precipitation from solution.

Orthoclase. *See* Felspar.

Orthogneiss. A gneiss of the igneous class.

Orthophyre. A medium-grained intermediate igneous rock allied to trachyte but containing well-shaped crystals of orthoclase; in minor intrusions.

Ouachitite. *See* Monchiquite.

Overthrust. *See* Thrust.

Paragneiss. A gneiss which originally was a sediment.

Pegmatite. A very coarsely crystallized igneous rock, most commonly like granite in mineralogical composition; usually in veins. The magma concerned was rich in fluxes (water, etc.) and so growth of large crystals was permitted.

Pelitic (Gk. *peles* = clay). Applied to metamorphosed argillaceous sediments (shales, etc.). When completely metamorphosed, the resulting rocks are mica-schists.

Peridotite. An ultrabasic igneous rock composed almost wholly of olivine; in dykes and also in larger masses.

Petrography. The science of the constitution, texture, etc. of all rocks. The term, *Petrology*, is more comprehensive, and includes the study of modes of occurrence, origin of magmas, etc.

Phenocryst. *See* Porphyritic.

Phlogopite. A light-coloured mica rich in magnesia; chiefly in metamorphosed dolomite.

Phyllite. A lowly metamorphosed argillaceous rock, characterized by a development of finely crystallized mica.

Picrite. An ultrabasic igneous rock composed of olivine and augite with some plagioclase felspar; in sills and dykes.

Pitchstone. A semi-glassy igneous rock with a greasy appearance and curved (conchoidal) fracture, usually acid in composition, often green in colour; chiefly in dykes and sills.

Plug. *See* Neck.

Plutonic. Applied to intrusive igneous rocks which usually occur in large masses and have crystallized slowly. The crystal-texture is in general coarse.

Porphyrite. An intermediate igneous rock consisting of porphyritic crystals of plagioclase felspar and a ferromagnesian mineral, usually hornblende, set in a fine-grained ground-mass of the same constituents; usually in dykes, the hypabyssal equivalent of diorite.

Porphyritic. Applied to a texture of crystalline rocks in which larger (porphyritic) crystals, termed *phenocrysts*, are set in a more finely crystalline or glassy ground-mass; especially characteristic of lavas and hypabyssal (minor) intrusions which have crystallized in two stages (i.e. before and after effusion or intrusion of the magma).

Prehnite. A pale green or colourless lime-rich hydrous mineral, often filling vesicles in lavas.

Psammitic (Gk. *psammos* = sand). Applied to metamorphosed arenaceous sediments (sandstone, arkose, etc.). When completely metamorphosed, the resulting rocks are granulites.

Pseudomorph. A mineral or mineral aggregate assuming the external form of a pre-existing mineral which it has replaced.

Pyrites. A brass-yellow mineral composed of iron sulphide, crystallizing in cubes, giving a greenish black streak when scratched with a penknife; frequently occurring in slates.

Pyroxene. A group of ferromagnesian minerals, characteristic of basic and many ultrabasic igneous rocks.

Quartz. A typically transparent mineral composed of silica (oxide of silicon), with irregular fracture. A constituent of acid, also in some intermediate, rare in basic igneous rocks.

Quartzite. A crystalline metamorphosed sandstone composed almost entirely of quartz, or with a fair proportion of felspar.

Quartz-porphyry. An acid rock consisting of porphyritic crystals chiefly of quartz and orthoclase felspar set in a fine-grained ground-mass of the same constituents; in dykes, sills, etc., the hypabyssal equivalent of granite.

Reversed Fault. *See* Fault.

Rhyolite. A fine-grained acid igneous rock, light in colour, with a streaky appearance due to flow of the partly crystallized magma; chiefly in lavas, the volcanic equivalent of granite.

Riebeckite. A mineral of the amphibole group, rich in soda, in colour dark bluish black.

Ring-dyke. A major intrusion with an arc or ring-shaped outcrop and with steep sides. Ring-dykes may occur one within another and so constitute an oval or circular plutonic complex, often about 6 miles in diameter. Two adjacent ring-dykes may be separated by a narrow belt (*screen*) of older rocks.

Sapphire. A very hard bright blue mineral composed of alumina; in highly metamorphosed argillaceous sediments.

Scapolite. A light-coloured mineral composed, like plagioclase felspar, of silicate of alumina with varying proportions of lime and soda; occurring as an alteration product of plagioclase felspar, and in metamorphosed calcareous rocks.

Schist. A class of metamorphic rocks with close-set planes of fissility (due to foliation); most commonly an altered sedimentary rock.

Screen. *See* Ring-dyke.

Secondary. Applied to minerals formed by the alteration of pre-existing minerals.

Sedimentary. Applied to rocks which originated as sediments, e.g. sandstone, shale, conglomerate.

Sericite. A variety of white mica in tiny flakes, commonly a decomposition product of potash felspar (orthoclase), a constituent of certain lowly metamorphosed schists.

Serpentine. A soft hydrous magnesia-rich rock, often green or red in colour, usually formed by the alteration of an olivine-rich rock, such as peridotite. Also a mineral.

Shale. A thinly bedded argillaceous rock.

Sill. A flat or gently inclined sheet of intrusive igneous rock, usually extending along the bedding of a sedimentary rock.

Sillimanite. A white mineral composed of silicate of alumina, in small closely packed crystal-needles; in highly metamorphosed argillaceous rocks.

Slide. *See* Thrust.

Spessartite. *See* Lamprophyre.

Spherulitic. Applied to a texture of igneous rocks in which finely crystallized elongate minerals (usually orthoclase felspar and quartz) occur in radially arranged aggregates (spherulites).

Spilite. A basic igneous rock allied to basalt but enriched in soda and so with albite instead of labradorite plagioclase felspar.

Spinel. A very hard dark mineral composed of magnesia and alumina; in small equidimensional crystals in ultrabasic igneous rocks and in metamorphosed dolomite, etc.

Staurolite. A brown-coloured mineral composed of silicate of alumina and iron often in twin crystals interpenetrated in the form of a cross; in highly metamorphosed argillaceous rocks.

Strike. The horizontal direction at right angles to the dip in bedded rocks.

Swarm. *See* Dyke-swarm.

Syenite. A coarsely crystallized intermediate igneous rock composed essentially of orthoclase felspar and hornblende; in major (plutonic) intrusions.

Syncline. An elongate trough-fold in bedded rocks which dip in directions inwards towards the axial plane of the fold, like a V.

Tachylyte. A basic glass, usually occurring as selvedges along the chilled margins of dykes.

Talc. A very soft magnesia-rich hydrous mineral, commonly green in colour, in thin plates resembling mica but with a greasy feel.

Teschenite. A medium-grained basic igneous rock allied to dolerite but containing analcime.

Tholeiite. A fine-grained basic igneous rock allied to basalt but with a proportion of glass dispersed in clots around or penetrating which crystals are arranged.

Thrust. A lowly inclined or flat plane of fracture (where not affected by folding) which is accompanied by displacement of the rocks on one side of the fracture relatively to those on the other. In some cases (N.W. Highlands) the rocks overlying the plane have been relatively displaced (*overthrust* of this volume), in other cases (Ballachulish) those underlying the plane (*slide* of this volume).

Tourmaline. A black or bluish black mineral composed of silicates of soda and alumina with iron or magnesia, in six-sided or triangular prisms; in pegmatite veins and pegmatite-enriched gneisses and schists of the Highlands.

Trachyte. A fine-grained pale or medium grey intermediate igneous rock, largely composed of lath-like crystals of orthoclase felspar; in lavas, and also dykes, the volcanic equivalent of syenite.

Tremolite. A pale or white variety of amphibole free of iron, often occurring in aggregates of fine needle-like crystals. Common in lime-rich metamorphic rocks.

Tuff. A medium- or fine-grained consolidated volcanic ash.

Ultrabasic rocks. Igneous rocks with a very low content of silica (less than 45 per cent.).

Unconformity. A rock-surface of a past geological time, due to erosion, upon which younger beds were deposited, is termed a plane of unconformity. The surface, in general, is parallel to the younger beds and makes an angle with the bedding of the older rocks. The younger beds are described as being unconformable with reference to the older.

Vesicle. *See* Amygdale.

Withamite. A red-coloured variety of epidote containing manganese; in veins and vesicles in lavas (Glen Coe).

Zeolites. A group of hydrous aluminous silicates of lime and/or soda found infilling vesicles of lavas, etc. (*see* Amygdale).

Zoisite. A pale brownish mineral allied to epidote, rich in lime and occurring in altered calcareous rocks.

SELECT LIST OF GEOLOGICAL MAPS OF THE WEST HIGHLANDS AND HEBRIDES

(Colour-printed editions of the Geological Survey)

Arranged with reference to Chapters concerned

	Maps on scale of	
Chapter	1 inch to 4 miles Sheet nos.	1 inch to 1 mile Sheet nos.
II	14	29,† 30,† 37
III	14, 16	Arran, 29†
IV	14, 16	Arran
V	13, 16	27, 28, 36
VI	13	19, 27, 28, 35, 36
VII	10,* 13	44, 45
VIII	10*	44, 53
IX	10*	44, 51, 52†
X	10,* 13	43, 44, 51
XI	7,† 10*	51, 60, 61,† 71
XII	10*	42 and 50, 51, 60
XIII	7†	70, 71, 80,† 81, 90*
XIV	4,‡ 5, 7†	Assynt, 81, 92
XV	4,‡ 7†	—

* In the Press.　　　† In preparation.　　　‡ Not yet prepared.

In addition to the above, a Geological Map of Scotland on the scale of 1 inch to 10 miles is in the Press, and maps of the Cuillin Hills and Western Red Hills of Skye are published, uncoloured, on the scale of 6 inches to a mile (Inverness-shire Sheet nos. 38, 39, 44 and 45).